S0-AVR-778

Wired for Music

Wired for Music

A Search for Health and Joy Through the Science of Sound

Adriana Barton

GREYSTONE BOOKS
Vancouver/Berkeley/London

References to health, medical interventions, traditional healing, and research in this book are not intended as medical advice, nor should they be regarded as a substitute for medical advice, professional expertise, or medical treatment. Although information provided in this book has been carefully considered by the author and the publisher, medical science is constantly changing and evolving. Therefore, all information in this book is provided without any warranty or guarantee on the part of the publisher or the author. Neither the author nor the publisher or their representatives shall bear any liability whatsoever for any personal injury, property damage, or financial losses.

22 23 24 25 26 5 4 3 2 1

Greystone Books Ltd.
greystonebooks.com

Cataloguing data available from Library and Archives Canada
ISBN 978-1-77164-554-6 (cloth)
ISBN 978-1-77164-555-3 (epub)

Editing by Lucy Kenward
Copy editing by Erin Parker
Proofreading by Jennifer Stewart
Index by Stephen Ullstrom
Text design by Belle Wuthrich

Printed and bound in Canada on FSC® certified paper at Friesens. The FSC® label means that materials used for the product have been responsibly sourced.

Greystone Books thanks the Canada Council for the Arts, the British Columbia Arts Council, the Province of British Columbia through the Book Publishing Tax Credit, and the Government of Canada for supporting our publishing activities.

Greystone Books gratefully acknowledges the xʷməθkʷəy̓əm (Musqueam), Sḵwx̱wú7mesh (Squamish), and səl̓ílwətaʔɬ (Tsleil-Waututh) peoples on whose land our Vancouver head office is located.

Contents

Prelude

I HELD A CELLO for the first time before a panel of stern-looking adults who peered down at me from a conference table, asking questions and taking notes. I was five years old—too young to realize this was an audition of sorts.

Funds were tight in our home in small-town Quebec, but my mother had caught wind of free violin lessons at the state-run conservatory a short drive away. Inside the drab building, the first teacher we met was a cellist with the Montreal Symphony Orchestra, on the hunt for new students. This was 1975, long before cello virtuoso Yo-Yo Ma became a household name, or a "Bond girl" in a Hollywood blockbuster used her cello case as a getaway sled. When the gray-haired teacher asked if I'd like to play the "cello," I had no clue what he meant. I heard "Jell-O" and nodded my head.

Before I knew it, I was sitting in front of the admissions committee on an upside-down wastepaper basket, the only seat small enough. The gray-haired man inspected my chubby fingers, stretched them apart, and announced they were strong enough for the job. Next, he handed me a quarter-sized plywood cello and showed me how to curl my right hand around the end of the bow. Then he told me to play. Tentatively, I drew the bow across the strings. *Screech!* I tried again. This time, the whisper of horsehair against metal became a wave of sound that rippled through me and bounced around the room. The next bow stroke triggered another wave of vibrations, another swell of delight. I remember searching for a polite way to describe this sensation to the adults in the room. "I can feel it in my bottom."

In those few moments, I discovered the vast, transformative power of sound—a shock of wonder I would never forget.

Introduction

RIFFLING THROUGH my first homeowner's insurance policy, I did the math: my most valuable possession was a cello I hadn't played in years. This took me aback because by then, in my mid-thirties, I couldn't even remember the last time I'd laid eyes on it.

The bulky instrument lay on its side, tucked behind the couch. I hauled the battered case to the middle of the room. *Did I really want to open it?* Unlatching the lid, I forced myself to look. My cello was covered in a layer of fine dust. As I gazed at its shapely form, another calculation hit me: I'd spent more hours of my life with this cello in my arms than with any lover—my husband included.

I plucked the A string. Slack and muffled by the case, it made a sickly note. When I picked up the bow, a plume of white horsehair fell across my wrist. The wiry hairs had detached from the tip. My throat tightened. I had never seen my bow like this. For a moment I thought about getting it rehaired, but decided there was no point.

I had given up making music.

Strapping the bow into the case, I closed the lid and stowed my cello back behind the couch. *What a waste*, I thought, for the umpteenth time.

For as long as I could remember, music had enchanted me. That's what music does. Some of us sing to Beyoncé when we're going through a rough patch, or collect vintage synthesizers to play '80s riffs from Duran Duran. Others have a thing for Balinese gamelan orchestras, Polish mazurkas, Swedish death metal.

Music moves us in mysterious ways. We get goosebumps, chills. A sudden urge to dance. Songs fill us with pleasure, revving up some of the same brain pathways stimulated by chocolate or sex.

The inability to enjoy music of any kind is so rare that brain specialists consider it a neurological condition: "Musical anhedonia" affects roughly 3 to 5 percent of us. People with this abnormality have glitches in the auditory-processing and reward systems of the brain. Everything from country to electronica leaves them cold.

The rest of us are wired for rhythm and song. Even if we lie perfectly still, music fires up the putamen, a nut-shaped structure at the base of the forebrain that helps regulate our motor movements. When music tickles our eardrums, our gray matter shimmies back.

Music's stimulating effect on the sentient jelly between our ears is more than just a geeky fact. During my twelve years as a journalist covering health at Canada's national newspaper, *The Globe and Mail*, I kept learning about the extraordinary ways music can support our health and well-being.

In Victoria, British Columbia, I interviewed a neurologic music therapist who explained that our brains process words differently when they are spoken versus sung. She had drawn fleeting responses from a brain-injured police officer, Ian Jordan, who had spent thirty years in a hospital bed, unable to communicate. When she sang the words, "Ian, lift your finger," he did precisely that.

Over the years, though, I spotted more and more articles and blog posts making laundry-list claims about "the power of music." All of a sudden, songs not only doubled as painkillers but could also supposedly make us smarter. Many of these reports smacked of pseudoscience (sorry, folks, but singing bowls cannot cure cancer) while others offered no data on what kind of music did the trick. Were they talking about clinical music therapy or cranking up Nicki Minaj in the car? Daily practice on an instrument, or playing the banjo on Sundays? I wanted the nitty-gritty, because in music, as with medicine, the details matter.

My need to separate the truth from the claptrap went beyond professional scrutiny. I was searching for a compelling reason to go anywhere near one of the biggest pain points in my life. Playing the cello had taught me that music really does work in mysterious ways. We can devote every spare minute to it and still miss out on its gifts.

I learned the hard way.

Before entering journalism, I spent seventeen years sawing away in a practice room, determined to become a professional cellist. I received instruction from cellists in the Cleveland Orchestra, Montreal Symphony, Canada's National Arts Centre Orchestra, and from international solo artist Antonio Lysy. In university ensembles, I performed at Roy Thomson Hall in Toronto, the National Arts Centre in Ottawa, and once at Carnegie Hall in New York.

Along the way, I developed soft-tissue injuries so severe it hurt to turn a doorknob. My worst injuries, though, were not physical. Drilled by my early teacher to make "each note a pearl," I became obsessive and perfectionist about my playing, to the point of anguish and despair.

MY BREAK WITH the cello coincided with dramatic advances in neuroscience that began in the 1990s, which U.S. President George Bush proclaimed the "Decade of the Brain." A new technique called functional magnetic resonance imaging (fMRI) allowed researchers to measure blood flow in the brain and see, in real time, how our gray and white matter respond to music. Neuroscientists discovered that rhythm and song fire up important brain regions, including the hippocampus (memory center), cerebellum (timing circuitry), and amygdala (emotional processing).

In many ways, the flurry of new findings validated a belief long held by Indigenous peoples: music is a strong elixir. The Irigwe people of central Nigeria use music and dance to heal owie dzio, "spoiled stomach," marked by abdominal pain and mental distress. In the lowlands of Siberia, Tuvan healers beat hand-held drums to make disease "fly away." Wishful thinking or not, healing strategies like these hold essential truths about our physiological makeup.

Our connection to rhythm and song is so profound that scientists have detected responses to music even in brain-injured people with severe disorders of consciousness. This is astonishing. While these patients may have only fleeting awareness of the outside world, their brain might still flicker to music.

Music therapists have leveraged this natural attunement for centuries, using music to treat ailments ranging from melancholy to small-motor injuries. In 1945, the U.S. War Department launched an ambitious music program for convalescing soldiers; within a year, nearly all of America's 122 veterans' hospitals offered music in operating rooms, psychiatric wards, and other facilities. An Army Air Forces private, Harold Rhodes, invented a bell-toned therapy instrument using aluminum tubing from wrecked B-17 bombers. Bedridden patients learned to play his "xylette," a lap-sized xylophone rigged to a piano keyboard. Rhodes went on to found an electric piano company, and in 1971, the Fender Rhodes Piano Bass gave its signature sound to the keyboard riffs in The Doors' hit "Riders on the Storm."

Decades later, universities worldwide began to rigorously test music's capacity to heal. Songs relieved pain in cancer patients. In surgical wards, music lowered anxiety as effectively as Valium. Familiar tunes revived memories, pulling the elderly out of the fog of dementia. Music offered preventative medicine, too. In a study aimed at reducing stress in people with early memory loss, listening to twelve minutes a day of relaxing music improved mood, sleep, and stress almost as well as the same amount of meditation training.

The more the case for music stacked up, the more I wondered what the new findings might mean for people like me, weighed down by musical baggage. I'd learned to avoid mentioning my first career at social events, because when I said "cello," people's eyes lit up. Then I'd have to answer the inevitable question: "Do you still play?" I didn't like having to explain that whenever I tried, my left shoulder seized and my forearms started to throb. Eyes would fill with pity. Then someone would say, "Oh, it's a shame you had to give it up."

It was. But I'm hardly unique. I've lost count of how many people have told me, in a wistful tone, that they'd love to play the cello, piano, or guitar. If only they had the talent, or hadn't quit when they were young.

Wade Davis, a *National Geographic* explorer, confessed to me that he can't sing. This "impediment," as he calls it, is "like a hole in my heart."

I was surprised to hear this. Now in his sixties, Davis has excelled as a writer, anthropologist, documentary filmmaker, and public speaker. Communication is his forte—yet he feels incapable of carrying a tune. Not even "Mary Had a Little Lamb."

Davis has a vague memory of someone making fun of him when he sang as a small child. Later, at age eleven, he was forced to take piano lessons every week in his teacher's Victorian-style parlor, where "she would put your fingers in clamps to make you do the chords." *Clamps?* They didn't hurt, he clarified, "but literally, she had these plastic things that you would put your hand in, and that's how she taught you to do the notes." Davis had just discovered The Beatles, but in his piano lessons, "there was no joy. She was making you play music that you hated, that had nothing to do with your life."

Over the years, his inability to express himself musically became a point of shame. Confronting this block, he said, "would change my entire being."

What's holding him back?

In many parts of the world, musical expression is considered a birthright. Some African and Indigenous languages don't have a specific word for "singer" or "musician"; it's a given that anyone who breathes can dance, drum, or sing. Music not only sprang from the human brain—it has the capacity to alter the structure and functioning of the brain itself. Aniruddh Patel, a music-cognition expert at Tufts University in Medford, Massachusetts, describes it as a "transformative technology of the mind."

How could someone like Davis be intimidated by this age-old source of connection and joy? For that matter, how could I? I wanted to get to the bottom of these hang-ups about music, this sense of missing out.

Scything through stacks of research, I decided that clinical music therapy should have its own separate book. This one focuses on the brain-altering effects of music that we can all tap in to. Along the way, I tackle questions that people kept asking me:

Are some kinds of music better for my brain?
Do I have to play an instrument to get the benefits?
Can music help my brain as I age?
Are trances real? Can music put me in a trance?
Can music harm as well as heal?

In organizing the material, I took cues from Abraham Maslow's "hierarchy of needs" (likely inspired, it turns out, by his experiences on a Blackfoot Reserve in the summer of 1938). Maslow proposed that psychological growth depends on fulfilling a series of needs, starting with food, water, shelter, and other survival basics. I looked at how musicality might have enhanced our species' survival. Next, I investigated how music can support everyday needs, from mental health and social relationships to brain development and healthy aging. Then I headed for the top of the Maslow pyramid—self-actualization—to explore music's role in creativity, spiritual growth, and the universal search for meaning.

This book is a blend of science and stories. I have changed the names of several people connected to me to protect privacy, and in certain places I have altered the chronology of my experiences to keep the narrative flowing. Other than these adjustments, all facts and events are accurate to the furthest of my ability to verify. (Full citations for all sources can be found in the notes.)

Throughout this book, I use "we" to convey broad themes in nations typically considered Westernized and wealthy.

I acknowledge that generalities do not reflect everyone's experience. While brevity is useful in writing, my goal is to reach anyone who longs to reclaim music as a strategy for well-being. (Nothing I have written, of course, can replace medical advice.)

This is not a how-to book. My aim is not to turn everyone into a musician. Instead, this is a why-to book: a passionate argument for following a yearning, however slight, to make more space for music in our lives. Because the truth is, even if we hear music every day, many of us don't realize just how potent rhythm and song can be.

In countries around the world, countless people make music for hours on end while others squint at their smartphones, scrolling their lives away. Social-media apps give us surges of dopamine, the brain chemical that prods us to seek pleasure. We keep scrolling despite growing proof that too much time on social media can be a fast track to loneliness, sleep problems, anxiety, and depression. Luckily for us, music also stimulates dopamine, while priming us for deeper meaning and connection to others.

From a time before memory, our early ancestors pulled music out of wood, seeds, animal skin, bone. Embedded in primeval rituals, and tucked inside our gray matter, are surprising answers to a simple question: How can music help us heal, and thrive?

— 1 —

Strings Attached

MY CELLO HAS RIBS of maple and a front of soft spruce, bonded together in Germany more than a century ago. Heavy and golden, it has many scars. A previous owner must have dragged it on hard surfaces, roughening the edges on both sides, and the base of its body shows a deep fissure in the wood grain, long since repaired.

I didn't choose this instrument.

When I was twelve, my teacher instructed my mother to find me a proper cello, saying I could no longer develop my sound on one of the conservatory's loaner models. My mom, never flush, paid a visit to the Ottawa studio of a Slovak luthier, Joseph Kun. With her purse filled with photographs of her artworks, she convinced the mustachioed craftsman to trade a cello for one of her paintings. (She has always maintained that Kun got the better deal.)

Of the three cellos on offer, I hankered for the dark glossy one, the color of tawny port. But my teacher, after playing them one by one, decided the golden one had the best sound for the price. And so I got the dinged-up cello.

Still, my fondness grew. My fingers knew every inch of its form: smooth on the front, dry in patches where the varnish had worn away. I learned just how hard I needed to push the wooden pegs to keep the metal strings from slipping out of tune, and how the sound of my cello would open up after the first hour of practice each day. In summer, its body would swell with humidity, like premenstrual bloat, pushing the wooden bridge so high that the strings sliced into my fingers. (Off to Mr. Kun, who carved a "summer bridge" for me.)

I kept my cello hanging from a hook on my bedroom wall so I could admire its faded beauty when I wasn't playing. Each morning at 6:30 sharp, I'd draw my bow across a chunk of rosin, the tree resin that strengthens the horsehair's grip. Then I'd place a tuning fork on the bridge of my cello to match its sound with my A string, feeling the sympathetic vibrations in the wooden panels as my cello hummed to life. Good morning. Time to practice our scales.

ON TUESDAYS AFTER SCHOOL, I waited for my cello lesson in the drab government building that housed the Conservatoire de musique du Québec. My lessons took place in a chilly studio with industrial carpeting, gray walls, a metal table, two chairs, and fluorescent lighting overhead. Other than the piano in the corner, it could have doubled as an interrogation room.

My cello teacher, André Mignault, had a French-Canadian accent and smoker's breath. At home, my stepfather liked to call him "Mr. Filet Mignon." But lessons with Monsieur Mignault were no joke.

Like most beginners on a stringed instrument, I struggled to play in tune. My teacher would pull apart the fingers of my left hand as far as they would go, forcing them to play whole tones on an instrument too big for my kindergarten hands. To this day, my left pinkie stretches a full inch farther than the right. My left shoulder, too, is permanently out of whack. But my teacher never mentioned my raised shoulder or tight grip on the bow. Instead, he would have me repeat the same note a dozen times before I could play the next, making micro-adjustments to the tuning, vibrato, or bow stroke.

Despite his good English, he often mixed up his H-sounds. "You must ERE with your HEARS," he'd say, pinching my earlobe so hard it burned. Then he'd scold me for the grimacing movements I made with my mouth, or for tapping my foot when I played. Too gauche for a classical musician.

Each lesson would proceed in slow-motion agony, with little music-making and few rewards.

Sometimes my mom would sit in and sketch. In one of her ink scribbles, both the cello's body and mine are cut off below the neck. Next to my face, the fingers of my left hand look like sausages, cramped at odd angles and grotesquely large. Mom would say she exaggerated my fingers to emphasize the powerful contortions at work, but I can hardly stand to look at this sketch. It makes playing the cello look painful and unnatural.

As he watched me tune my cello, M. Mignault asked the same question every week, every month, every year: "How much did you practice?" By the time I was ten, the answer was two hours a day. "Not enough," he'd say. "At your age I was practicing at least three hours, maybe four—that's the only way you're ever going to make it."

He never asked if I wanted to make it as a cellist. This was a given. M. Mignault not only performed as a cellist in the Montreal Symphony Orchestra but was also an instructor at the Université de Montréal. Unless a child aimed high enough, she was wasting his time.

I was no child prodigy, but I won top marks at his prodding. From the start, M. Mignault told me he would never give me compliments, "because then you'll stop practicing." Even when I became the youngest student ever to pass the second of the conservatory's four performance levels, he kept his word.

Still, it wasn't all pain, no gain. Near the end of each lesson, the piano accompanist would slip into the room. Trim and soft-spoken, Denise Pépin always flashed me a smile. M. Mignault would still say "arrête" (stop) multiple times, but he couldn't force a pianist of her caliber to keep repeating every phrase. As she played the opening chords of my sonata or concerto, I'd feel a bubbling excitement. Then I'd try to match her rhythms, gestures, and feel. For a few minutes, before the next "arrête," Madame Pépin and I would gallop away.

Moments like these kept me going for eight years with M. Mignault. But in my diary, at age twelve, I confided my deepest wish: "I want to be good, but most of all I want to have pleasure playing the cello." Sadly, this goal never fit in my regimen as a professional musician-in-training.

By the age of thirteen, I'd spent more than 340 hours of my life in a room with M. Mignault. We'd become like an embittered old couple, constantly finding fault with each other. The more I chafed at his demands for extra practicing, higher scores, the more he nagged and browbeat me. Just before I turned fourteen, I left him for another teacher, a cellist with the National Arts Centre Orchestra in Ottawa. I don't remember giving an explanation or thanking M. Mignault for everything he'd taught me. I had grown to hate him.

My new teacher, David Hutchenreuther (known in our family as "Hutch"), had a kinder, warmer style. We'd talk about the meaning of the sonatas and concertos I played and the different tone qualities I could pull from the cello. He helped me think about how I'd like the piece to sound. During my three years with him, I won a giant trophy in my first competition and joined a youth orchestra in Ottawa, a welcome change from the conservatory's stiff ensembles.

Up until then, I had been driven by the need to please M. Mignault, not to mention my mom. But with Hutch, I began to see a future for myself as a professional musician.

FOR MY SIXTEENTH BIRTHDAY, my mother took me to see Yo-Yo Ma perform Haydn's *Cello Concerto in D*, a frothy yet technically demanding piece. Ma played with an ease that left me goggle-eyed. *Would I ever learn to play like that?* Afterwards, Mom had another surprise: a chance to meet the thirty-year-old virtuoso backstage.

Dashing in his tuxedo, Ma had large wire-rimmed glasses ('80s style) and dark hair that brushed his shirt collar, slightly unkempt. In the backstage hallway, a cluster of fans gathered around him, but I was the youngest of the lot. My mom pulled out a copy of Ma's latest album and told him I was studying the cello. He smiled at me and asked, "What are you playing?"

The Saint-Saëns concerto, I replied. On the album, he wrote in bold letters, "To Adriana Happy Birthday!!! + good wishes for S.S. etc. etc. YYM." But that wasn't all.

"Would you like to try my cello?" he asked. *Would I!* He ushered me into his dressing room, shutting the door on his other admirers. Then he handed me his Stradivari, carved by the legendary Italian craftsman nearly three hundred years before. I could hardly breathe. The cello was the color of amber and surprisingly light. I was terrified of dropping it. Tentatively I pulled a few notes of the Saint-Saëns out of this magnificent instrument, a voluptuous wooden lung. But instead of playing more, I asked a technical question about the opening line—though I already knew the answer. Then my time with Ma was up.

I thanked him and left the room, kicking myself. By this time, I had more than a decade of cello under my belt. Why couldn't I savor the moment and play my heart out?

Music wasn't something I did on the spot. I had only performed in formal concerts, competitions, and annual juried exams. When guests in our home begged me to "play us a song," I always balked. I didn't know any songs. All I knew were sonatas, études, and concertos. And my early training had taught me that nothing short of flawless could ever be good enough.

IN HIGH SCHOOL, I spent my lunch hours in the library working on assignments to save extra time for practicing at home. Whenever I finished early, though, I'd treat myself to some chill time with the library's subscription to a British classical music magazine launched in 1890. Leafing through *The Strad* one day, I spotted a full-page ad for the Cleveland Institute of Music showing a man cradling his cello as if it were a child. I read every word. He was Stephen Geber, principal cellist of the Cleveland Orchestra, routinely listed among the top-five orchestras in the world. Although I had never heard him play, his look of deep reverence made an impression on me. I vowed to become his student.

No time to waste. At age sixteen, I already felt over the hill; cellists much younger than me had made their Carnegie Hall debuts. I pulled together an application as fast as I could, and a week after learning what an SAT was, I wrote the standard exams required for U.S. college admission. Four months later, my mom drove me five hundred miles from Ottawa for my audition at the Cleveland Institute of Music

As we approached the outskirts of Cleveland, derelict homes and abandoned warehouses sprawled as far as I could see. *What was I getting myself into?* By the time we pulled up to the Howard

Johnson motel, I was beat. I hauled my cello out of the back seat—and slammed the car door on the third digit of my fingering hand. Mom went through the possibilities. Was it nerves? Self-sabotage? My finger swelled to twice its size. As I iced it, I was furious with myself. My audition was shot.

The next morning, we made our way to the Cleveland Institute of Music, a welcome sight after our drive through the Rust Belt city. The school was part of Case Western Reserve University, laid out on a leafy campus long before the steel industry went bust. In the audition room, half a dozen instructors greeted me, including Mr. Geber. He was a tall man with a dimpled chin and warm eyes. When I told him about my injured finger, he smiled and told me to do my best.

I was nervous. With no pianist to ground me, I botched the opening of my concerto and then rushed the arpeggios, despite having practiced this piece a million times. *Why wouldn't my fingers cooperate?* I thanked the admissions committee and slunk out of the room.

Mom had been listening in the hall. Her eyes shone. "You did it!" No, I said, shaking my head. "It was all over the place."

"But you had the sound, the depth," she countered. "I'm sure they heard that."

Six weeks later, a crisp white envelope arrived in the mail. I sat on the stairs, clutching it. It was too thin for an acceptance package. With a sigh, I told myself maybe this rejection was for the best. I wouldn't have to scramble to finish high school early and could spend more time with Hutch. I could pace myself for once. In my gut, I felt a flicker of relief.

"We are pleased to confirm your acceptance to the Bachelor of Music program in Cello Performance," the embossed sheet of paper declared. Forms for a partial scholarship, a paid work-study position, and college registration would follow. I stared at the letter, rereading it several times. *Was this really happening?* Stephen Geber taught fewer than ten students each year, from bachelor's level to doctorate. My heart beat faster. He had chosen me.

THAT SPRING, my mother drove me to the National Arts Centre in Ottawa for me to compete for a $2,200 prize. I needed the cash for college, but at sixteen, I was up against musicians aged eighteen to twenty-four. The only piece I'd practiced enough was a concerto by the French composer Édouard Lalo, a moody tour de force. The cello starts with a cryptic unaccompanied theme before unleashing a furious *allegro maestoso.* With a pianist filling in for an entire orchestra, I'd have to summon enough drama for a monumental work.

Other than the judges in the front row and my mom in the back, the recital hall was empty. After taking a bow, I sat on the chair in the middle of the stage and turned to the accompanist, signaling her to begin. As I dug my bow into the first low G, I tried to hear the dark tormented theme in my head, and imagine the orchestra punctuating each flourish. But in that cavernous hall, with eleven pairs of eyes scrutinizing my every move, every phrase I played seemed to flounder like sailboats in a gale. I fumbled through the first movement, trying to salvage the shreds. Then it was over. "Thank you," one of the judges said.

I fled the scene. Later, as I sniffled in my bedroom, one of the judges phoned my mother to bring me back for the next round. She coaxed me back into the car, back on stage. I can barely recall the final round. My only vivid memory is of how unnatural my playing sounded to me. As undeserving as I felt, though, I won a bursary that day.

A month later, a marble-shaped cyst appeared in my right wrist, followed by another in the left. But I didn't think to ask myself if my body was trying to tell me something. I had gotten into a prestigious college that attracted graduates of Juilliard. I wasn't about to give up.

IN THE OLDEN DAYS, physicians would take a Bible and smash it down on a ganglion cyst, erupting the fluid-filled mass by force. But this method had fallen out of favor, my doctor said, because more often than not, the cyst came back. Bulging out of a joint or the lining of a tendon, ganglions tend to keep growing unless they are surgically removed, and even then, they often recur. Seeing my horrified look, the doctor gave me an alternative: immobilize the area.

My first lesson in Cleveland started with my telling Mr. Geber that I couldn't play. "Don't worry," he reassured me. The renowned Cleveland Clinic had just launched a program for treating musicians and dancers with performance injuries (more common than I'd known). For the next three months, I did hand exercises and wore a splint. Although the ganglions shrank, I was also diagnosed with tendinitis, a repetitive strain injury that became so severe I could barely comb my hair.

I resolved to start practicing again at the start of the winter term, no matter what. *Get a grip*, I told myself. Tylenol helped keep the pain down and I learned to play with swollen forearms, along with the stress of soft-tissue injuries that kept coming back.

My new teacher advised me to take it easy. "Just do what you can." For me, though, "take it easy" didn't compute. Envious, I watched his other students blossom by the week. I felt so behind.

Mr. Geber had me play my scales, études, and sonatas at a quarter speed. I was to mimic his every bow stroke, variation in vibrato, and musical inflection, down to his every breath. While his method clearly worked, something about it troubled me. Hutch had encouraged me to think about the history and intentions behind a composition and consider the different ideas it could convey. When I tried to talk about musical interpretation with Mr. Geber, though, he cut me short: "For now, play as I play. Later on, you can develop your style." But I had my own ideas, my own emotions. Why did they have to wait?

For the first time since I was a kid, I couldn't stick to a practice schedule. I'd cram for my lesson, promising myself I'd do better next week, but each time Mr. Geber didn't chide me for showing up unprepared, I felt worse. I couldn't face the cello the next day, or the day after that. Then I'd find myself practicing furiously at the last minute, inflaming my tendinitis as the lesson loomed.

Guilt festered, too. My parents had remortgaged our house to cover my expenses. They had three other children to support. And there I was, screwing up.

The next fall, my teacher told me I'd need to make up for lost time to fulfill the school's requirements. Although he didn't want me to strain my wrists, I'd have to prepare for two juried exams instead of one, and show the same amount of progress between them that would normally take a year. Driven by the fear of losing my scholarship, I whipped my practice habits into shape enough to pull off both performance exams. But this achievement gave me no pride. In my mind, I had merely caught up.

At the end of the term, I took a bus home to Ottawa without a word to my teacher about the summer session he had arranged for me. The house was empty. My stepfather was at work and my mother and younger siblings had left early for the annual visit to see our grandparents in Nova Scotia. I left my cello and suitcase downstairs and headed straight to my room. For the next two days, I didn't eat or talk to anyone. I slept and slept.

Then, on the third day, a knock on my bedroom door. "Mr. Geber is on the phone," my stepfather said. "He wishes to speak with you." I picked up the receiver and mustered a greeting.

"What's going on, Adriana?" Mr. Geber asked. I told him I wasn't coming back to Cleveland. My parents couldn't afford to keep paying for college. "Don't worry about that," he said. "We can sort that out."

"But I'm so far behind. I don't want to waste your time."

Mr. Geber reminded me that I'd done well on my first performance exam, "and the second one was acceptable under the circumstances," he said. "You've made a lot of progress this year."

I didn't know how to tell him that nothing about Cleveland felt right—from the crime-ridden city to his teaching approach. I told him, again, that I wasn't coming back.

There was a pause at the other end of the phone. Mr. Geber repeated that I was making a bad decision, and for the first time I heard frustration in his voice. He added that he'd invested two years in me. Then he hung up.

THE SUMMER CRAWLED BY as I waitressed at a diner in a shopping mall, slinging plates of hot dogs and gravy-coated fries. I decided I should at least finish my degree. But I couldn't stomach going back to Cleveland, even if Mr. Geber would have me. On a day off work, I phoned McGill University in Montreal and learned that the cellist Antonio Lysy, an international solo artist, was joining the faculty and looking for students. I took a two-hour bus ride to Montreal for an audition. Although I hadn't practiced all summer, three weeks before the start of the term I was offered a spot.

Unlike Cleveland, Montreal dazzled me with its Paris-in-North-America flair. I loved the saturated colors of the sugar maples in fall, the jauntily dressed bon vivants on Boulevard Saint-Laurent, and the crumbling Victorian apartment buildings in my new neighborhood, Plateau Mont-Royal.

In the McGill music department, a violinist and flute player invited me to join them as classical musicians for hire, in a group we called Trio con Brio. Dressed in white blouses and long black skirts, we played in the metro for spare change in between gigs that included lavish receptions at the Four Seasons hotel. At wedding jobs, we gritted our teeth when a betrothed couple requested a sappy number like Andrew Lloyd Webber's "Memories" from *Cats*. For a hundred bucks an hour each, though, we'd play anything.

At McGill, I redoubled my efforts after my Cleveland defeat, practicing four or five hours a day on top of orchestra and quartet rehearsals, private lessons, and masterclasses. My compulsion to practice took masochistic forms. On a rare weekend getaway, I trudged up a hill in the Laurentian Mountains with my cello strapped on my back because the weight of my instrument was less than the burden I would feel if I missed a practice, even on holiday.

To me, this was normal. Many of my peers shackled themselves to their instruments for the love of music and thrill of performing. McGill's closet-sized practice studios started filling up at 7 a.m., and the din of scales and orchestral licks kept echoing through the walls until around ten at night. Student lore had it that a performance major's workload rivaled that of a med student, a comparison so plausible to us, no one bothered to verify it.

I lived for the high points, like the time my university orchestra performed at Carnegie Hall. Although it was hardly a solo debut, at age twenty I'd become living proof of the adage "How do you get to Carnegie Hall? Practice, practice, practice."

During the run-through, I could have pinched myself. Had I really made it to the hallowed hall graced by Tchaikovsky in 1891, the night these doors opened for the first time? While it hardly seemed possible, I might even be playing in the same spot Paul McCartney stood in 1964, the year The Beatles took New York by storm. All through the rehearsal, I kept gawking at the elliptical ceiling and the gilded columns jutting from creamy walls. *So fancy.*

That night, each of the hall's 2,804 seats was full. Buried in the orchestra, I had no nerves, only excitement, as we performed an Antonio Vivaldi piece with a full choir, followed by a cacophonous symphony by Erich Wolfgang Korngold. Although we didn't get a standing ovation, the rumbling applause hyped us up.

Minutes later, we hopped across the street to the Park Central Hotel, where the entire orchestra was staying for the night. Our conductor, Timothy Vernon, had invited everyone to an after-party in his suite, which got so rowdy that the hotel security booted us to an empty ballroom on the thirteenth floor. The drunken revelry continued long after I turned in.

The next morning, Vernon slurred a congratulatory speech whilst chugging a can of beer on the bus. Hair of the dog. Playing Carnegie Hall was a peak moment for all of us, and I still have the glossy program with my name printed inside. As spectacular as it was, though, this performance wasn't the one I remember most from my Montreal years.

INSIDE A CHAPEL on a dark December afternoon, the light of three hundred candles shone on faces taut with grief. The only sound was the shuffling of feet. I was twenty-one, sitting with my cello to the side of the crowd, waiting for the signal to play. Rubbing my hands together, I tried to warm them in the chilly air. My hand trembled as I raised my bow.

Montrealers had gathered to mark unthinkable loss. Two years before, a misogynist gunman had opened fire at an engineering school across town. On December 6, 1989, fourteen women were killed.

Gazing at the stricken faces in the chapel, I had no idea how a trembling cellist could comfort anyone, but it was too late to back out. So, I took a deep breath and played the first note of the prelude from Bach's *Cello Suite no. 2 in D Minor*. I was shaking so badly I could feel my bow bouncing off the strings—the cellist's version of stuttering. It sounded like someone trying to sing while a pair of hands closed around their throat. Seconds later, as my muscle memory kicked in, jittery notes gave way to fluid arpeggios. Bach's steady rhythms took it from there.

The solemn prelude hummed from the wooden body of my cello. Minor chords rang out. In the flickering candlelight, I thought of the fourteen women who hadn't lived to see this day. Women my age, ambitious, whip-smart. Beloved by their families. As Bach's spiraling cadences echoed in the rafters, the air grew thick with sadness, anger, and fear. The intensity grew almost too much to bear. Moments later, the hymn-like melody tapered to a series of long, steady chords. Hush.

That night, a clip from the vigil played on the national news. I saw the cello and the faces rapt in sounds. Strains of Bach, tender yet steely, had absorbed the terrorizing thoughts. For those brief moments, music had invited the gathering to feel and find solace. A kind of healing.

But not for me.

THE NIGHT BEFORE my solo in the chapel, I had spiraled into panic, terrified of not measuring up to the grave occasion. What if I dropped my bow? What if I froze onstage? Although I performed several times a month, I was convinced this time I was done for. I hid under my bedcovers, weeping with dread. The awful feeling wasn't just stage fright—I could cope with that. This was a dread that had seeped into my marrow, a penetrating conviction of my own inadequacies. If I wasn't good enough at the cello, I was nothing.

After the vigil, shaky and depressed, I found myself a therapist named Nansea. She recommended a book, *You Can Heal Your Life*, by Louise Hay, an American cult figure who claimed to have cured her terminal cancer through loving affirmations and self-forgiveness. On Nansea's advice, I wrote page after page of lines like "I am a good person. I am a talented musician. I deserve to be alive." But no matter how many notebooks I filled, I couldn't write my way out of paralyzing self-doubt.

Near the end of my Bachelor of Music degree, I had another flare-up of tendinitis. It started with a dull ache, the telltale sign, and progressed to an angry throbbing that swelled and radiated in both arms. *Why does this keep happening to me?* Once again, I had to get a doctor's note to delay my performance exam, ashamed of my weakness, betrayed by my body. Finally, at age twenty-two, I quit. I hid my cello in a corner of my apartment, latched in a case that could hold a small child. I felt guilt, grief. It was as if my twin had died.

MANY PEOPLE DESCRIBE music as a "universal language," a tuneful lingua franca. But this metaphor, penned in 1835 by the American poet Henry Wadsworth Longfellow, assumes the same music speaks to us all. I have my doubts.

Chinese opera and American chill pop have as much in common as Korean and Haida. The Tsimane people of Bolivia rate consonant and dissonant sounds as equally pleasing, and the hunter-gatherers of the Congo don't necessarily perceive music in minor keys as sad.

Although most of us can identify another culture's dance tune or lullaby, we don't share a common musical language. Instead, we share a common hunger to interact with rhythm and song, often in ways that are intensely personal. Every time we choose an instrument or select a track from a music player, we reveal something about where we come from. Who we are.

Without the cello, who was I?

Music was my only identity. I had never done anything else. I was no good at sports or relationships and had no other passions or dreams.

But I still needed to make rent. Montreal had sunk into an economic depression and I took the only job I could find, as a receptionist at CHOM-FM radio station, "Montreal's Home for Classic Rock." There I was, a classical-music geek, greeting rock stars I'd never heard of in the waiting area: Good afternoon, Alice Cooper. Hello, Meatloaf.

My crash course in pop music came with perks, including free tickets to see the Beastie Boys perform an entire set while bouncing on a trampoline. I watched the crowd egging them on. The band was so cheeky, so brash. Around the same time, I dated a guy my age from the newsroom and realized I was mostly attracted to his job—a blend of learning about the world and performing on air. Months later, I entered the graduate journalism program at Concordia University, steps away from the chapel where I'd performed two years before.

As I cut my teeth as a reporter, I began to let loose in summer festivals and nightclubs, joining the mosh pit at ska and hip-hop shows. Live bands were a revelation. Never once had classical music given me the urge to dance, or even bop in my seat.

Encountering so many different ways of making music gave rise to an unsettling thought: maybe, despite seventeen years of cello training, I had never really learned what music was all about.

— 2 —
The Music Instinct

ON A CLAMMY NIGHT in Montreal, Dave, the guy sharing my futon, rolled towards me and said, "You're not a real musician." I was twenty-three, partway through journalism school. Dave had never heard me sing, or play a single note, but the cello case in the corner of my bedroom told him everything he needed to know.

"You never made your own music. You never jammed." His breath smelled of cigarettes and the nine-dollar wine we'd shared from the convenience store around the corner. "All you did was rehash someone else's songs." He paused to scratch the stubble under his chin. "That's not music," he said, "that's parasitic."

I stared at the water stains on the ceiling, eyes welling. I wanted to tell him he was wrong about me, that I was just as musical as he was. But I couldn't forget all the times I'd felt like a trained monkey, playing by rote. *Widen your vibrato here. Insert dramatic pause there. Was that music?* I mumbled something about musical interpretation being its own artform, but even I wasn't convinced.

Dave, you see, was a real musician. A brooding drummer who test-drove his songs on street corners with his snare drum and hi-hat. I'd met him at a friend's party in a smoky loft and then followed him and his Gretsch kit to the grimiest bars in town, thrilling to the sound of his pulsing tom-toms and bass—his sweat, physicality, and total abandon.

I had never played music like that. Compared to the feral rumpus he made, the concertos I'd memorized belonged in a museum dedicated to men in curly wigs and tights.

Dave preferred army pants and a battered leather jacket, the uniform of a punk mystic. He practiced drums every morning, then tai chi, twisting his wiry frame into moves like Golden Cock Stands on One Leg. Later, he'd drink and get high and scramble words into songs that gave grammar the finger but often rhymed. Once, he wrote a love song for me: "Lost in Zen."

He got the "lost" part right.

We called it quits after a road trip through the Rockies in a lumbering Dodge sedan—a bargain at four hundred bucks—with his dog in the back and my cello in the trunk. Once we hit Vancouver, I stayed put, scrounging a living as a freelance writer. Dave took off to California, but his words stayed with me: "not a real musician."

What did that even mean?

Some days I had to fight the urge to track him down and make him take it back. I might not have written my own songs, but my cello playing had moved people, made them laugh, cry. That must count for something.

Since before I could read words, music had inscribed itself in my muscle memory. To scratch "musician" from my tagline was to erase a part of myself. The next guy to question my musical identity got an earful. By this time, though, I didn't need a therapist to tell me these men were simply reflecting my own thoughts. What's a musician who doesn't play music? A has-been. A fake.

This refrain kept tormenting me. Like Dave, I grew up thinking of musical talent as a gift: either you have it, or you don't. My fear of being in the "have not" category kept prodding me throughout my cello training. Maybe, with enough practice, no one would notice what I didn't have.

It would take years before I understood that musicality involves a lot more than pro skills at an instrument or performing on stage. Although some of us have stronger musical skills than others, the myth of musical talent has been largely debunked. As the neuroscientist Daniel Levitin, author of *This Is Your Brain on Music*, explains, there's no such thing as a "music gene" or a "center in the brain that Stevie Wonder has that nobody else does."

What if capacities for rhythm and song weren't as rare as we think? What if, instead, musicality was the norm?

LIKE OTHER ANIMALS, humans evolved in a world thrumming with sounds—whistling winds, cawing crows, chirping squirrels, burping frogs. At some point, our hairy ancestors got in on the action, banging sticks together, blowing through wooden tubes, trilling at the top of their lungs. What purpose did this raucous instinct serve? Birds and whales "sing" to communicate, using signals understood by their kind. Humans, though, communicate mainly through language. Music, if you think about it, has no clear raison d'être.

Yet every human society ever studied has dedicated outrageous amounts of energy to songs of celebration, rhythms for working the fields, dances to mourn the dead. Gary Tomlinson, a musicologist at Yale University, points out that a human culture devoid of music "simply doesn't exist." Based on the best evidence we have, "music is as ubiquitous a capacity among humans as is language."

His words struck me as a thinly veiled dig at Steven Pinker, the best-selling author and cognitive psychologist at Harvard University. Pinker enraged musicologists worldwide in his 1997 book *How the Mind Works*, dismissing music as nothing more than "auditory cheesecake."

In Pinker's view, music merely piggybacked on the mental frameworks that evolved to support functions such as language. Music, he reasoned, has no biological importance. It does not extend our life span, help us predict threats, or enhance procreation (assuming one ignores the bump-and-grind on dance floors). In fact, music could vanish from our species "and the rest of our lifestyle would be virtually unchanged."

Pinker didn't get the last word, though. Critics have argued that he and other scholars went about solving the riddle of music in entirely the wrong way. From Darwin onward, great minds pondered questions like "How did music help humanity survive?" And when they failed to find clear answers, they relegated music to an acquired skill—a product of nurture, not nature.

Tomlinson took a radically different tack. Instead of asking what purpose music in its current form might possibly serve, he asked: What kind of cognitive capacities are needed to make music? How might they have developed? Others examined the fruits of our musical labors, while Tomlinson looked for clues in the metaphorical soils and weather conditions needed to make them grow.

His work came together in a groundbreaking book, *A Million Years of Music: The Emergence of Human Modernity*. Drawing from cognitive neuroscience, evolutionary biology, and the arcane field of paleo-musicology, Tomlinson traced the origins of music to a time before language, before human culture—before our ancestors could conceive of a distinct self. The building blocks of music, he wrote, helped shape the modern brain. One reviewer described Tomlinson's work as the "ultimate rebuttal" of Pinker's "cheesecake" theory, recasting music "as an essential part of human identity, rivaling speech."

Not everyone thinks of music as part of their identity. Even so, the theory of musicality as intrinsic to our species offers the most convincing argument I've seen for why so many of us are drawn to music—crave it, even.

As I waded through Tomlinson's dense academic text, my thoughts drifted towards my own origins—before I became a promising cellist and then "not a real musician." I started digging through family memorabilia, searching for clues about my earliest days. Then something dawned on me: without music, I might not be here at all.

MY FATHER, A UKRAINIAN named Yuri, grew up in a Polish village behind the Iron Curtain. He was pale and slight, a Soviet math whiz. At seventeen, he immigrated to Canada to study engineering at the University of Toronto. There, he ended up in the same boarding house as my mother, Susan, a visual-arts student from Nova Scotia who had landed a summer job at an art gallery in the big city. In the downstairs sitting room, the boarding house had an upright piano. One evening, as Susan plinked away, Yuri, blue-eyed and raven-haired, sidled up to her and said, "You play beautifully."

She practically swooned over the keyboard. "It was the foreign accent," she told me, "and the sound of his voice."

Mom described Yuri as a passionate aesthete with a yen for classical music, computer science, yoga, and Zen Buddhism. In a series of articles from the mid-1960s, the *Toronto Star* chronicled my father's overland journey from England to India, where he communed with yogis who hadn't left their caves in years. Mom married him in Turkey, on his return trip.

While expecting me, my parents lived in a one-room cabin in Carlsbad Springs, a rural area not far from Ottawa. Mom drew water from a brook and wrote letters to my grandmother on birch bark, describing the cabin's cheery orange-and-yellow curtains, wooden rocking chair, and Depression-era woodstove.

My parents owned just a handful of records in this cabin, where I came into the world during my older sister's nap. I was born—and likely conceived, my mother said—to the chiseled strains of Glenn Gould's 1955 recording of Bach's *Goldberg Variations*. (She still has the vinyl tucked away somewhere.)

My father died of colorectal cancer at twenty-nine, when I was a year old. I grew up knowing very little about him. But the day I arrived earthside, he gave me two names: Adriana, the *A*s pronounced "ah," as in "apple," followed by Natasha. "Ah! Ah!" he told my mother. "She must have Ah!"

IF A STEALTHY RESEARCHER had snuck into the cabin with an EEG machine to record brain activity, they might have discovered something remarkable about the newborn brain (though not mine in particular). Electrodes attached to my fuzzy scalp would pick up electrical signals timed to the beat of the *Goldberg Variations* spinning on the record player. Even if I were napping, my brain would register the basic pulse of the toccata or fugue. I wouldn't need to come from a musical family to do this, because the human brain is wired to find patterns in musical rhythm. Researchers demonstrated this trait by playing drumbeats to sleeping infants hooked up to EEG machines. When they took out the downbeat—the strong pulse that makes us tap our feet—the newborn brains could predict where it should be (just like adults in the same study). This blew me away. The ability to perceive a musical beat, observed the 2009 study, is "functional at birth."

If this finding is correct (and so far, it remains unchallenged), at a neurological level, we all have rhythm. This means even those with "two left feet" on the dance floor have a hope of learning a few moves if they put in the time (just like Willard, the dance-deprived teenager in the movie *Footloose*).

Our brains attune to rhythms through entrainment, the tendency for vibrating objects to lock into phase. Christian Huygens, Dutch inventor of the pendulum clock, discovered entrainment in 1665, noting that his clocks would eventually swing together, in "odd sympathy." Like cuckoo clocks, our brainwaves tend to oscillate in time with any steady pulse we hear.

This brainwave entrainment, which develops in early childhood, allows us to tap our feet to a beat, from reggae to cha-cha. While it might take practice to get the coordination down, our brains are primed for musical rhythms.

How did humans get their groove? Henkjan Honing, a professor of music cognition at the University of Amsterdam, explores this question in his remarkable book *The Evolving Animal Orchestra: In Search of What Makes Us Musical*. We might think our innate sense of timing comes from the thump-thump of a mother's heartbeat. That's not it. All mammals hear their mother's heartbeat in the womb, but precious few can follow a beat. When researchers adapted the EEG experiment on newborns for a study of rhesus macaques (a type of monkey native to Southeast Asia), they found no sign that macaque brains could detect a musical pulse.

Naturally, scientists assumed that beat perception was a uniquely human trait. Then along came Snowball, a sulphur-crested cockatoo, whose dance moves to the Backstreet Boys became a YouTube sensation. For music-cognition specialist Aniruddh Patel, speaking to *The New York Times* in 2010, watching a bird entraining to a pop song was like seeing a dog reading a newspaper out loud. "My jaw hit the floor."

Patel rushed to the shelter in Indiana where Snowball lived to see if this bird could keep the beat when the same song was sped up or slowed down. To the researcher's astonishment, Snowball nailed it for nine out of eleven tempos. The boogying cockatoo became the first documented case of an untrained animal that could match a musical beat. (Snowball later hit the big time bopping to "The Piña Colada Song" in a TV ad for Taco Bell.)

Researchers wracked their brains to figure out what this cockatoo had that rhesus macaques, who share 93 percent of their DNA with humans, do not. Birds, like humans, have a knack for mimicking sounds. And phrases like "Polly wanna cracker?" have rhythm as well as pitch. Beat-matching, Patel speculated, must come with vocal learning.

This made perfect sense—until Ronan, a California sea lion, bucked the theory. In exchange for bites of herring, the peppy sea mammal would bob her head up and down to the beat of a metronome and switch tempos on a dime. But sea lions aren't vocal learners. Although they growl and cough to communicate with their young, they don't have the vocal systems needed to mimic sounds. That didn't stop Ronan from "headbanging" in perfect time to the disco hit "Boogie Wonderland."

The vocal-learning theory got another blow in 2014, when Japanese researchers observed a chimpanzee swaying to a beat while its mother practiced matching a pulse on a piano keyboard. (Bribing with apples helped.) Compared to birds or humans, chimpanzees show limited vocal learning. Yet these captive chimps could more or less follow a beat.

Clearly, vocal learning doesn't fully explain beat perception. And the human monopoly on music is a myth. In 2016, the

New Scientist reported that wild gorillas in the Republic of the Congo make "little food songs" at mealtimes. An animal keeper, Ali Vella-Irving, made a similar observation at the Toronto Zoo: "Each gorilla has its own voice. You can really tell who's singing," she said. "And if it's their favorite food, they sing louder."

In my eyes, none of these findings diminishes the wonder of human musicality. On the contrary. From singing whales to beat-matching budgies, musical animals give hints that the neurological underpinnings of music might be older and more broadly shared than we ever imagined.

The human brain has strong two-way connections between our auditory (hearing) cortex and motor-control (movement) center. Without this link, we couldn't learn to dance or play guitar nearly as well as we do. Chimps have the same neural network but theirs is feeble. Somewhere down the line, human ancestors stumbled on ways to sharpen it for reasons that chimps, our closest relatives, didn't have.

We separated from our last common ancestor roughly six million years ago. Something momentous in our evolution must have happened after the split.

MUSIC, LIKE OUR SPECIES, has origins in East Africa.

Along the highway between the Serengeti grasslands and Tanzania's Ngorongoro Crater—a meadowy oasis teeming with lions, zebras, elephants, and wildebeest—two colossal skulls of concrete mark the turnoff to the "Cradle of Humankind." Signs lead to Oldupai Gorge, source of the hominin skulls and Paleolithic tools that reveal the genius of our ancestors in deep time. (The famous gorge is also known as Olduvai; both spellings are derived from a Maasai word meaning "place of the wild sisal.")

The museum at this active archaeological site has a room dedicated to a cast of the fossilized trail discovered in 1978 by

Mary Leakey at nearby Laetoli. I remember seeing these footprints in the pages of *National Geographic* when I was a kid. They looked like tracks I'd make on a wet beach, except these prints were 3.6 million years old. Imprinted by feet hardly different from our own, the Laetoli trail proved that our ancestors walked upright, with a human-like stride, at least two million years before the massive growth spurt in the hominin brain.

In these footprints, Anton Killin, an evolutionary theorist, sees traces of rhythm.

The trail shows two sets of footprints, side by side. But a closer look reveals a third individual, clearly stepping in the larger footprints. Following in another's footsteps requires adjusting one's stride and synchronizing to a different gait and walking speed. Other apes don't do this. For our small-brained ancestor, *Australopithecus afarensis*, the ability to match strides along a seventy-print trail amounted to a giant leap.

With every step, the early walkers must have heard the squish-squish of feet sinking into wet volcanic ash. Did these squishy sounds, combined with muscle movements, reinforce the auditory-motor connection in the brain—the one that's much stronger in bipedal humans than in chimps?

The idea of innate rhythm embedded in primeval footprints delighted me. But ancient neural connections leave no fossil record. They're impossible to prove.

Paleolithic tools offer more tangible clues. Tomlinson, the Yale musicologist, believes the building blocks of music emerged from the rhythmic chipping of rock on rock. *Homo erectus* began chiseling diamond-shaped hand-axes about 1.7 million years ago. Generation after generation gained this skill by mimicking other knappers (chippers of stone). Over time, unconscious mimicry led to voluntary motor control. It took focus—and several hundred hours of practice—to learn to knap a hand-axe. Neurons that fire together, wire together. Thus, as our ancestors sharpened stones, tool-making sharpened the brain.

Granite, when struck, sounds different from limestone or flint. Learning to tune in to these differences, Killin wrote, must have guided knappers in their choice of materials, enhancing their skill set. The knappers' children, raised with these percussive sounds, might have practiced their rhythmic timing with other little hominins during sessions of "lithic sound play." (The original "rock music.")

Next came fire. *Homo erectus* learned to harness flames at least 790,000 years ago, and possibly earlier. Cooked foods yielded more energy. Nourished by roasted tubers, hippopotamus, elephant, and boar, early brains doubled in size.

Fire added leisure time. Instead of spending up to half of their waking hours chewing raw foods, as other primates do, *Homo erectus* had time to spruce up their shelters and lounge around the fire after dark. Bored and restless, they might have bonded by synchronizing their voices and body movements, said Killin, "rehearsing" precursors of song and dance. Cooking fires, over centuries, fanned the flames of human imagination.

LIKE A MOTH TO FLAME, my mother was drawn to music and adventure.

Widowed at age twenty-six by my father's death, Mom hitched a ride from Ottawa to Vancouver with two toddlers in tow. Within weeks, she fell in with a band of ragtag musicians, enthralled by their counterculture ideals and wildly spontaneous sounds. I was in diapers when she got cozy with the Pied Piper of the group, a tall and bushy-haired Californian named Michael Fles. A minor beat poet, he had partied with Allen Ginsberg and fought U.S. censors to publish the first excerpts of William S. Burroughs's *Naked Lunch*. Some years later, Michael's stint in California's underground marijuana industry got him slapped with a drug charge. Reluctant to face the music, he hightailed to Canada.

In Vancouver, Michael worked the system, applying for a government grant to fund "spontaneous music workshops" for children with special needs. He got the idea from Christopher Tree, the musician he'd followed to Woodstock the year before, in the summer of 1969. Tree, a musical wizard, had opened each day of the festival with a cacophony of Tibetan temple gongs. "I was his roadie and manager," said Michael, now in his eighties, when I tracked him down by phone. "He's the guy who introduced me to the gongs."

Tree's spirit-of-the-moment philosophy guided Michael and his shaggy band of followers in their musical happenings in and around Vancouver. Half a dozen of them, Mom included, would show up at a center for children with disabilities, unroll a Persian carpet, and lay out an assortment of bamboo flutes, whistles, hand drums, wind chimes, shakers, and car parts dangling from strings. In grainy footage I found online, children with Down syndrome bang on pot lids and xylophones, looking mesmerized as Michael pounds the gongs and Johnnie, my mom's next boyfriend, plays the clarinet. (For a second, I glimpse a long-haired woman who looks like my mother, tapping a drum.)

Mom often talked about these hippie jam sessions when I was a kid, but I never clued in that I was part of them, too. "Of course you were," she said recently. She brought me and my older sister to every event, letting us play with any bell or rattle that tickled our fancy. (Dave, the punk drummer, was wrong. I jammed before I could talk.)

Mom didn't stick around Vancouver for long because the men in her life had itchy feet. Together with Johnnie, we piled into a truck and followed Michael, the Pied Piper, all the way to Mexico. In the mountains of Chiapas, Michael and Johnnie hung out with local musicians while Mom tried her hand at Mayan pottery.

My earliest memory is of a bonfire crackling in daylight as a long line of women carry earthen pots to the flames. Mom said this scene fits a time when I was three, in the pottery village of Amatenango del Valle. We slept in an abandoned stable, much to the amusement of the Mayan villagers. *Gringos locos*, they must have thought. A black-and-white photo shows me in pigtails and a handwoven dress, sitting in the dirt. My chubby hands clutch a pottery flute shaped like a child with a round face and curly hair. The clay figure is me. I am blowing into the hidden mouthpiece, filling myself with sound.

I have no memory of playing this ocarina, or of the two musicians who inspired my mom to make flutes of clay. All I have is a faded photo of Johnnie, bearded and ringlet-haired, with his arm around my mom and sister, and me on his lap. We look like a family, beaming in the tropical sun. My sister and I called him "Daddy," Mom said. Weeks after this photo was taken, though, she left him for Russell, my future stepfather, a man-about-town she'd met in Ottawa before our journey out west.

We said goodbye to Johnnie at an airport in Guatemala. I'll never know what music, if any, I absorbed from him and Michael. Still, their anything-goes ethos left a legacy for others.

Back in Vancouver, two members of the "children's spontaneous music workshops" regrouped in 1976 to launch Canada's first training program in music therapy, at Capilano College (now Capilano University). One of the founders, Nancy McMaster, said the program never strayed from the values of spontaneity and free expression rooted in "Woodstock, the '60s, and some hitchhikers." The budding music therapists encouraged children with special needs, many of them nonverbal, to tune in to their own musical impulses moment to moment, and communicate in ways that transcended words.

While their approach was radical at the time, it was hardly counterintuitive. The earliest humans—also nonverbal—might have made music much like this.

SILVER-TONGUED SCHOLARS INSIST that language is the essence of our humanity, the dividing line between us and other animals. But this bias ignores the deeply embodied, imitative, and emotional aspects of human cognition. Long before we had anything like language or music, our nonverbal ancestors learned to think and communicate in intricate ways.

Half a million years ago, in a balmy period in southern England, *Homo heidelbergensis* hunted rhinoceros and megaloceros (giant deer), leaving a cache of bones and stone tools at Boxgrove in West Sussex. Archaeologists noted that each carcass was skillfully butchered: "Fillet steaks were sliced from the spine, and the bones were smashed to get out the marrow."

Feeding the clan depended on group choreography and constant communication. Warriors kept long-toothed scavengers at bay. Knappers made hand-axes for stripping flesh, while underlings gathered branches for the fire. To divvy up the spoils and reinforce the chains of command without a brawl, the leaders would have hollered, pounded the ground, beaten their chests. Anatomical changes helped them make themselves clear. These hunters had lost the air sacs around the larynx that allow other primates to "hoot-pant" without hyperventilating. Without these air sacs, the Boxgrove band could vocalize with close to human finesse.

Changes in the outer and middle ear helped our ancestors tune in to each other's voices like never before. Over time, their brains began to pick out distinct pitches from all the sounds around them. This floors me: the separate tones we hear in music and speech are mental perceptions—all in our heads. In nature, there's no such thing as "middle C." Every sound vibrates with many frequencies (oscillations per second) at once, but our brains perceive the lowest frequency as a distinct note, ignoring the higher frequencies, called harmonics or overtones. Other primates perceive pitch differently than humans do. But we hear distinct notes without even thinking about it.

At some point, our ancestors had both pitch perception and rhythm, but before we could dream up true language or music, we had to learn how to combine smaller components into a whole. Just as language is made up of words and sentences, music combines notes and silences into patterns of rhythm and melody. This ability to piece things together, step by step, doesn't show up until about 250,000 years ago, around the rise of *Homo sapiens*. Tool makers attached wooden handles to chiseled blades using strips of leather, manufacturing each component separately before putting them together. Complex thought processes flourished in the prehistoric assembly line.

The building blocks for language and music developed over eons on a parallel track. Then they parted ways. Language evolved for symbolic thinking (the sound "apple" stands for a particular fruit), whereas music sent direct signals to the emotion and command centers of the brain. Gentle sounds calmed us. Aggressive ones pumped us up. Brain scanning has confirmed that contrary to Pinker's "cheesecake" theory, we don't need our language system to process music.

No one can pinpoint the precise birthdate of music on the timeline leading to "behavioral modernity," the dawn of the modern mind. Starting about fifty thousand years ago, growing capacities for symbolic thinking and innovation yielded striking cave paintings, knives and fishhooks crafted from antler, and shiny beads carved of shell. But music, in a form we'd recognize today, almost certainly came earlier.

During the last ice age, small bands of sapiens huddled in the caves of Germany's Swabian Alps, playing flutes fashioned from mammoth ivory and vulture bone. These flutes look like penny whistles, with finger holes carefully spaced to play tunes in five notes, similar to the pentatonic scales found in Scottish jigs or Javanese gamelan music. (On a replica of the vulture-bone flute, one experimental archaeologist managed to play "The Star-Spangled Banner.") To carve such sophisticated flutes, ice-age humans needed a feel for melody. Long before making these instruments, carbon-dated at roughly forty thousand years, our ancestors must have improvised with their voices and drummed on hollow logs.

Music's longevity in our species points to a biological purpose: Songs might have made living in groups more bearable, increasing our chances of survival. Dancing enhanced Paleo romance, along with procreation, while tunes that soothed a baby's cries lowered the risk of predator attacks.

Researchers at Harvard University's Music Lab have discovered that babies will calm down to lullabies in any language, from Hopi to Polynesian. The soothing doesn't come from hearing a parent's voice or a familiar musical style. Babies are wired for rhythm and song.

Unborn babies respond to the contours of melodies and, after birth, can distinguish between unfamiliar songs and tunes they heard in the womb. (Playing classical music for unborn babies does not make them smarter, though. Samuel Mehr, director of Harvard's Music Lab, calls this the "pop-sci myth that refuses to die." Whether the music is classical, hip-hop, or grunge, the only benefit is that an unborn baby might remember it later on.)

Only at six months old do infants use their pitch and beat perception to recognize components of language. In infant development, wrote Honing, the Dutch music-cognition specialist, "musicality precedes both music and language."

Our capacities for music evolved over millennia, in tandem with the modern brain. Pitch perception, beat perception, and the ability to find meaning in patterns of sound are part of what makes us human. Remarkably, our auditory cortex can rewire itself to process touch as well as sound, allowing deaf people and those hard of hearing to dance, play instruments, and enjoy music. The renowned percussionist Evelyn Glennie, profoundly deaf since age twelve, performs shoeless to allow her feet to feel the vibrations from the floor. The human body, she told *The Globe and Mail*, is "like a huge ear."

We are musical creatures. Arguably, the most musical on Earth.

LEARNING ABOUT the prehistoric origins of music filled me with awe but also left me perplexed. If humans have innate capacities for rhythm and song, then why do so many people doubt they have a musical bone in their body? In Western industrial societies, up to a quarter of adults will describe themselves as unmusical. This disconnect from our musical wiring made no sense until I thought of other human instincts that have withered, like hunting, spear-throwing, and even breastfeeding.

The ability to nurse our young defines us as mammals, yet my friend Emily, a nurse, spends days and nights in a maternity ward helping new mothers reconnect with this trait. Early attempts may end in tears because many women doubt their ability to feed their own babies. How could a mammal lose touch with this instinct?

In 1865, a German chemist got the bright idea to replace breast milk with a formula called Liebig's Soluble Food for Babies. When brands such as Similac ("similar to lactation") began marketing their powders to hospitals and pediatricians, infant formula took off. By 1971, a mere 10 percent of American babies were breastfed for at least four months.

Since then, strategies to revive breastfeeding have turned a bonding ritual into a mechanical procedure: "Squeeze your breast like a hamburger! Check the latch!" My friend Emily takes a different approach, reminding new mothers that they come from a long line of successful breastfeeders. "I tell them, 'Put your baby on your chest, enjoy being together, and trust that you and your baby will figure this out, even if you need a little help.'"

In less than a century, formula makers came close to cutting off our connection to the milk of life. Could a similar disconnect happen in people starved of a musically rich environment?

Most children can learn to hum in tune and dance to the beat, grinning all the while. Anyone who doubts this can check out videos of preschoolers wiggling to music in Mexico, Cuba, or Senegal. In many places, to this day, music is inseparable from work, play, religion, and medicine. While social environments go a long way in nurturing these skills, children are naturally musical. Deep down, so are adults.

Some scholars describe music as "central for human well-being," which might seem like a stretch. After all, we can live without songs, just as we can go without breast milk, vegetables, or forest walks. But basic survival is the lot of a broiler chicken. To thrive, we cannot ignore our biology.

We evolved in forests and on savannahs, living in tight-knit social groups. Our neurobiology hasn't caught up with urban environments that have us dodging strangers on crowded streets clanging with traffic sounds. In a study of attention and memory, even just looking at pictures of nature, compared to urban scenes, improved people's scores by 20 percent. The same goes for music. Even just listening to a few minutes of slow-paced rhythms coaxes our gray matter into calming down our breathing and pulse. Music brings us back to another kind of knowing—a way of sensing and feeling, as Wade Davis puts it, "what it means to be human and alive."

On a primal level, we understand this. That's why it hurts to be told "you must be tone deaf" or "you have no rhythm," especially if we believe it ourselves. It's like being told we cannot feel normal emotions such as sadness or love. Attacks on our capacities for music amount to attacks on our humanity.

When the punk drummer called me "not a real musician," he stabbed a nerve because part of me already felt deficient. Even at peak cello, I could sense something missing, something that had nothing to do with talent or skill. Now I understand why. By the age of five, I'd gone from the most free-flowing hippie healing vibes in North America to the strictest music training this side of the Atlantic. Somewhere along the way, I had lost something wild, something precious. I had no idea how to get it back.

IF I'D KNOWN SOONER about the deep roots of music in my early life, and in the human brain, I might have had something to hold on to in my late twenties when I tried to reinvent myself as a cellist.

At the time, I struggled to pay the bills. Besides my freelance gig writing for a *Reader's Digest* coffee-table book series, I cleaned houses and worked as an extra on film sets, wearing everything from zombie makeup to a torpedo bra for the role of a '50s waitress. When I came home from set, often in the middle of the night, my cello case in the corner would confront me. *What are you doing with your life?*

Hidden in the case was a gift from Dave, a pickup for my cello that could transmit sound through an amplifier. Even as he denigrated my classical training, he was savvy enough to know his punk band might get a leg up with a cello in the mix. The throaty strains of cello music had spread to pop bands and TV commercials. Cellos were hot.

I never did play in Dave's band, but one day I scrounged together fifteen bucks to place an ad in Vancouver's entertainment weekly: "Classical cellist seeks pop, jazz, or rock musicians. Let's jam." I tinkered with the wording on and off for hours, wanting to sound confident yet casual. But every word I wrote sounded fake. *What if no one answered my ad?* Or just as nerve-wracking: *What if someone did?*

I got a call from a bass player in a cabaret band. Over the phone, I came clean: "I haven't played in a while," I admitted. "And I've never jammed."

"That's cool," he said. "Bring your cello to my place."

I found myself carrying my cello through an alley that reeked of urine as I made my way to the back of an old warehouse in Vancouver's Downtown Eastside. A scrawny figure opened the door. He had a beaked nose and sleepy eyes. *Typical hungover musician*, I thought. I sat down on a vinyl chair with a gash in the seat and he watched as I rosined my bow. My shoulders tensed. I recognized the look. Gaga over the cello.

He started riffing on a simple bass line while I plucked a note here and there, not really knowing what to play next. He looked up and nodded as I began drawing long notes with my bow, liking how they wrapped around the deep purrs of the bass. I could feel my shoulders settling down. Phew. He didn't seem to expect much. After about twenty minutes, my forearms began to twinge. "That's all I can do for now," I said. He nodded slowly and smiled. "Sure. Let's do it again soon."

A month later, he showed up with his bass on the doorstep of my one-room apartment. Pale and sunken-eyed, he looked rougher than I remembered. We picked up where we left off, but I worried my neighbors would complain about his portable amp thudding through the walls. I couldn't focus.

Over glasses of water, the bass player opened up. "Sorry it took me so long to call you back. I was kind of out of it."

"Why's that?"

He paused. "Heroin," he said, quickly adding, "I don't shoot up. I only inhale."

"Oh." I glanced away, trying not to look shocked. He seemed like a nice guy and I didn't want to judge. For all I knew, smoking heroin was the norm in his music crowd. I thought of Billie Holiday, John Coltrane, Steve Tyler, Kurt Cobain... Still, the prim classical cellist in me recoiled. I was afraid of hard drugs, and instead of finding empathy for him, I beat myself up with the thought that the only one who wanted to make music with me was on smack.

The next musician to answer my ad was an alt-country-blues singer who had just recorded her first single. She showed up wearing a vintage dress and retro heels, a getup fit for Nashville's Grand Ole Opry. I liked her husky alto and the quirky ballads she'd penned, but she only played her own songs, waiting for me to chime in. I didn't know any blues riffs, so I suggested we try notes that neither of us had played before. She huffed. Desperate to make it work, I offered to practice blues scales for next time. She never called me back.

I canceled my ad. Outside the classical world, I didn't know where I belonged. I couldn't afford lessons to learn a new musical language. And my arms still hurt.

Months later, I met a piano teacher in her early fifties who misted up when she heard about my musical past. "You don't have to play the cello," she said, offering to work on songs together, free of charge. "You can sing instead."

I came prepared with lines I'd scribbled after my split with Dave: "Sometimes it's sunny / But mostly it pours / When I leave my world / To live in yours."

"Cheesy, huh?" I said sheepishly. She didn't mind: "Lots of songs are cheesy."

She told me the basics of songwriting—the verse, the bridge, the chorus. Then she played different chords on her guitar until we found the ones I liked. "How about a D major chord?" she said, strumming on the downstroke. I shook my head. "Then let's try D dominant 7th." *That's it!* She was so patient, so kind. But we had barely written a song or two when she moved out of town, two ferry rides away. The little faith I'd had in myself left with her.

Without a musical guide, I had no defense against the inner whispers telling me I had nothing to offer musically besides a few lines of bad poetry. I was a no-good cellist, a trained monkey. Not a real musician. Once again, I gave up.

Despite every failure, though, and every toxic thought, I couldn't stop feeling music's gravitational pull.

— 3 —
Groove, Interrupted

AT A BEACHSIDE FESTIVAL in Vancouver, I wandered through the crowds for hours, listening to different bands thud-thudding on stage. I was thirty, still reeling from another blown-up relationship and moldering as a tourism-magazine editor when what I really wanted was to write. By late afternoon, the air was muggy and every speck of shade was taken. I headed for the exit. Just then, I heard music with a bounce to it—a catchy, hyperactive quality that drew me back in.

The band, Cascabulho, mashed together drums and electric bass with folk songs from Brazil's Northeast. When I tracked them down after the show, they told me about a music scene called mangue, named for the mangrove swamps around the port city of Recife. Six months later, I followed them to Brazil. (Not for a fling or a stint as a groupie, but with a hope that some of their buoyant energy might rub off.)

"People think of Brazil as this dreamy land of rainforests, samba, and micro bikinis," the band leader, Kleber Magrão, told me. "But here in the Northeast, the land is arid. Conditions are tough. We have our own music, our own traditions, but even people from Rio don't know about them."

The ragged metropolis of Recife took me aback. Beyond the compact colonial district, settled in the 1530s by the Portuguese, the city of more than three million was chockablock with crumbling apartment buildings and hand-built favelas. The sight of street children bathing in ditches stopped me in my tracks. I didn't want to stare, but there were so many of them, so small. I couldn't look away.

For bands like Cascabulho, music was a manifesto, a drumming of pride in the hardscrabble nerve of this place. Kleber guided me to rehearsals and evening gigs featuring groups with names like Mestre Ambrósio. While their songs had depth and a cosmopolitan grit, the snarling guitars sounded routine to me next to the beats they'd borrowed from the nearby fishing towns and cane fields.

One sweltering morning, we took three buses to reach a favela, its dirt streets pungent with open sewage. Toddlers in flip-flops and women with colorful headwraps sang in a solemn procession that wound its way to a dusty courtyard. There, an elderly man in a soccer jersey hammered out beats on a cowbell, whipping everyone into a frenzied call-and-response. I bobbed along, swept up by the rhythms.

Later, Kleber would explain that this ceremony honored the "Kings of Congo," laborers once "crowned" as leaders among slaves. Brazil abolished slavery in 1888. At first, I didn't understand why such a ceremony would be kept alive, despite its ties to African systems of social order and not just to the Portuguese crown. Then I realized that these songs might help people remember what their forebears had endured, and how far they'd come. Even small children knew all the words.

Wherever we went, people sang one minute and drummed the next, dancing all the while. A trio of young women studying sciences at the state university showed me their chops on a two-sided drum called the alfaia, played with thin sticks that made a heavy booming sound. I envied how easy they made it look. But here, no one batted an eye if someone played a smidgen before the beat or sang a little off-key. It was all part of the musical texture.

Along with African rhythms, I heard wheezy accordions, an indigenous bamboo flute called a pífano, and the rabeca, an archaic fiddle wildly popular in these parts. Hipsters played tambourines while rappers dueled with embolada, a tongue-twisting street poetry. The names of the folk styles had their own bump and sway: côco, caboclinho, maracatú, forró (for all).

When Kleber described frevo, an acrobatic dance with tiny umbrellas, I told him I'd like to see that. He laughed, and said, "No one does frevo at this time of year. Now, we do côco." His musical traditions enchanted me. Even in a sprawling concrete city, rhythms could rise and fall with the seasons.

Before I left Brazil, Kleber convinced me to load my suitcase with local CDs. Then he handed me a cassette tape with bass tracks from the songs I liked most. "You can practice them on your cello and play along." Smiling, I reminded him that I didn't play anymore. "Why not?" he asked, and I could see he didn't get it. His friends noodled around on their instruments all the time, at beach parties and in city squares, between swigs of crisp Bohemia beer. (No worries about noise by-laws.)

After two weeks in Brazil, I didn't understand what was holding me back either. I liked the music, and Kleber's bass tracks would be easy for me to learn. Why couldn't I see myself grooving along? My reluctance made more sense when I got back home.

WHERE I COME FROM, countless people don't dare sing except in the shower. Others swear they couldn't carry a beat to save their life. Many are intimidated by music—alienated, even. Kleber would be shocked.

The neuroscientist Daniel Levitin sums it up: "There's this strange phenomenon that people are afraid to express themselves musically." They'll insist they've never been good at music or joke they're doing others a favor by keeping their mouths shut. No one would say they can't chat at a cocktail party because they're no Martin Luther King. "But we do say, 'I'm no Aretha Franklin, so I'm not going to sing.'"

Nearly everyone is born with the same musical wiring as Kleber and his crew. What, then, is tripping us up?

Chances are we're not "tone deaf," more accurately called amusia. Less than 2 percent of people are unable to hear the difference between, say, a white key on a piano and the black key next to it. Problems distinguishing between pitches involve congenital abnormalities in the brain and may be partly hereditary. But amusia hardly accounts for the roughly 17 to 25 percent of adults in Western industrial societies who swear they have a "tin ear."

Most of us can feel the beat in music, too. Difficulties with beat perception are so rare that the first case didn't show up in the scientific literature until 2011, when researchers at McGill University described a graduate student, Mathieu, as "beat deaf." Mathieu could neither move in time to a beat nor tell if someone else was dancing in time. Unlike Mathieu, however, most people who think they have no rhythm are simply uncoordinated or haven't practiced moving to a beat. A lack of beat perception is even rarer than tonal amusia.

Here's the crux: the deficits we call "tone deafness" or "left-footedness" are largely cultural.

That said, people have good reasons for refusing to dance or sing. Researchers have uncovered an array of painful emotions—from anxiety to shame—in those who merely mouth the words during national anthems or "Happy Birthday." There's even a term for this: "selective mutism for singing."

Picture a small child with a quavering voice, struggling to hit the notes. The day before the school pageant, the teacher tells the child, in front of the whole class, to sing softly or not at all. "Just for the performance," the teacher adds, as if this makes it better. Fighting back tears, the child decides (consciously or not) to avoid future humiliation by never singing again.

Moments like these deliver a "fatal blow" to a child's musical self-image, wrote the late Steven Demorest, a professor of music education at Northwestern University. His work has shown that children who have a negative view of themselves as singers are far less likely to join in music of any kind.

The saddest part: most of these children have as much musical potential as the next child. Singing and auditory skills, just like reading, develop at a different pace in different kids. But all too often, teachers act like talent scouts, weeding out the "bad" singers from the "good." Instead of nurturing the weaker ones, they abandon them, reinforcing the strong cultural belief that some people just can't sing. This becomes a self-fulfilling prophecy (though not set in stone; see chapter 9).

How does a society end up with adults who tell children not to sing, and classical music that has no groove—no dynamic push-pull feel or "swing"?

THE MORE I LEARNED about our species' innate musicality, the more my culture's approach to music struck me as bizarre. The tight control over musical talent goes against our evolutionary history. Each of us comes from a long line of music-makers, even if we've never sung or played an instrument ourselves. And yet, inhibitions run deep.

I dusted off my history books, determined to pinpoint the source of these hang-ups.

The ancient Greeks believed everyone could and should have a knack for music, most often blended with poetry and dance. Writing on fragile papyrus scrolls, they described music as a source of divine wisdom from the Muses, the nine goddesses of literature, science, and the arts. But playing an instrument wasn't just for a-Muse-ment.

In the sixth century BCE, Pythagoras reportedly prescribed certain harp songs as "uppers" to sharpen mental clarity and others as "downers" for relaxation at the end of the day. Meanwhile, this guru-like figure convinced his followers that planetary bodies moved according to the same mathematical ratios found in musical scales.

More than a century later, taking cues from Pythagoras, Plato reasoned that music could tune the body and soul to the "harmony of the spheres." The Muses gave humanity the gift of attunement "not for mindless pleasure," Plato wrote, "but as an aid to bringing our soul-circuit, when it has got out of tune, into order and harmony." For the Greeks, music doubled as mind-body medicine. More than that, it made celestial bodies go round.

Partway through the Roman Empire, however, music was in for a massive upheaval.

Pagan Rome was a rollicking place: Trumpets, reed pipes, and tambourines pumped up the volume in chariot races, carnival trains, and gladiator fights. Day in and day out, the Romans sacrificed goats and sang hymns to their pantheon of gods, including Bacchus, god of fertility and grapes. Bacchus beckoned worshippers to drink wine and let loose to the "howlings" of drums, noted a disapproving witness, until they entered a state of ekstasis (Greek for being outside one's self). But as the mobs joined in, these cathartic rites devolved into drunken riots and sexual free-for-alls.

Against this backdrop, Christianity took hold. The Church Fathers realized that to triumph over paganism, the new religion would need to sever all ties to dirty dancing. So, they targeted the pulse of Roman life: rhythm.

I realize this statement might sound overblown, but in 195 CE, Clement of Alexandria made himself clear: "The devious spells of syncopated tunes," he wrote, "corrupt morals by their sensual and affected style, and insidiously inflame the passions."

While a harp was permitted to accompany hymns at home, the Church Fathers implored the faithful to abandon their horns, bugles, and drums. Panpipes and flutes, wrote Clement, were "better suited to beasts." Recognizing music's power, the priests encouraged the chanting of hymns to open hearts to God's teachings. But they urged believers to reject all dancing and elaborate singing, lest their souls be tainted by "diabolical choruses."

Now we jump ahead to the fourth century CE, when Christianity became the official state religion across the Roman Empire. Although villagers must have strummed at house parties on the sly, under the watchful eye of the priests, any pleasure in music was suspect, even in church. Saint Augustine, in roughly 400 CE, confessed to being "enthralled" by the sound of hymns and feared that if he was more moved by music than the text itself, he was "sinning criminally."

This guilt-ridden mindset would eventually give rise to the most austere music I've ever heard: Gregorian chant. Sung in unison—all the same note—Gregorian chant moves mostly stepwise in small leaps, using only the notes you'd find on the white keys of a piano. This style gathered momentum as an oral tradition until the ninth century when, in written form, it became the official music of the Catholic Church for nearly a thousand years. In this solemn chanting of the Latin liturgy, many listeners find transcendent beauty. Even so, this is music stripped to the bone—no harmony, musical instruments, or distinct beat. The priests, true to their aim, had replaced the trance-inducing rhythms of pagan Rome with music that barely had a pulse.

Ancient history, you might say. Or is it? We take for granted the musical chords we hear in pop tunes and the cadences that make a Beethoven symphony feel "finished." Nevertheless, the entire harmonic system of Western European music developed through the Church, step by step.

After Gregorian chant, only gradually did the Church allow harmony (different notes sung together) to trickle back in. First came the octave along with fourths and fifths, followed later by thirds and sixths.

Church composers avoided dissonant sounds, especially the Devil's tritone, known by its Middle Ages nickname, Diabolus in musica, "the Devil in music." (In the devilish *Simpsons* theme song, the first two notes form a tritone.) This combination of three whole tones was considered so jarring and difficult to lead into a pleasing harmony that medieval monks were taught not to use it. As the musicologist John Deathridge explained in an interview with the BBC: "You can read into that a theological ban in the guise of a technical ban."

In university, I studied counterpoint, the system of relationships between harmonies and musical voices later used by Vivaldi and Bach. Writing even a short piece of counterpoint was like playing chess, because each era had different restrictions on how one note could move to the next. In a sixteenth-century crackdown, for example, the Church's Council of Trent warned that harmonies "must not give empty pleasure to the ear" and ordered priests to "banish from church" all sounds "that are luscious or impure."

For many of us, the rules embedded in this system continue to shape how we learn to play instruments and think about music—even if we never listen to Beethoven or Bach. As of 2020, this antiquated genre accounted for just 1 percent of all music consumed in the U.S. (compared to 28 percent for hip-hop and R&B and 19 percent for rock). Yet classical music clings to its pedestal.

Recently, a new acquaintance told me about the electro-trance music he listens to in the car. "I don't know if it's the rhythm or what, but it helps me focus." In the next breath, though, he denigrated his own tunes: "I should probably be listening to Mozart or something like that. I'll bet classical would be better for my brain."

DESPITE THE CHURCH'S sweeping influence, we've always had pockets of boisterous singing, drumming, and swing-yer-partner culture. In the old world, polka, flamenco, and Celtic tunes are still going strong. Across the pond, folks in Cape Breton, Nova Scotia, grow up with fiddle music. Quebecois families gather in the kitchen playing spoons. Inuit communities are reviving throat singing, while Indigenous children pick up drum songs that have survived colonial rule.

Farther south, bluegrass and country draw fans from miles around. And to this day, from Justin Timberlake to Avril Lavigne, dozens of pop stars get their start in church choirs.

Then there are families like mine. My sister took violin lessons while I played cello, but none of us made music together. Although my mother had nine years of piano as a child, she tickled the ivories just once or twice a year. To be fair, even if she'd had the inclination, she seldom had the time. A year after I began cello lessons, my little brother was born with congenital heart abnormalities that left him battling pneumonia and the constant risk of heart failure. When I was ten years old, my parents took him to England for a surgery that had never been done in Canada, leaving me and my older sister in Ottawa with friends. For nearly two months, I didn't know if I'd ever see my little brother again. He came home with a thick red scar down his chest, at age five, a medical miracle.

In the years my parents drove him back and forth to the hospital, the conservatory kept me and my sister busy. Practicing for our performance exams and orchestra rehearsals was demanding enough, so we never bothered learning extra parts to play in the living room together "just for fun."

Our family didn't have a pop-music tradition either, other than the scratched-up Beatles records my British stepfather, Russell, often played. I loved "Sgt. Pepper's Lonely Hearts Club Band" and could sing every word of "Lucy in the Sky With Diamonds" by heart. But the thought of teaching myself the cello part from "Eleanor Rigby" never crossed my mind. Now I understand why. My sister and I didn't play our instruments. We *worked* them.

WHENEVER I COMPLAINED about my first teacher, M. Mignault, my mother would empathize with me. Then she'd defend him. He might be strict, she'd say, "but that's the nature of the system that your teacher grew up in—the French system." I didn't know what she meant.

Now, looking back, I can't call him a cruel man. In an obituary from 2018, I saw laugh lines etched in his eighty-six-year-old face. When I looked into his background, I was surprised to find write-ups on my first teachers in the *Canadian Encyclopedia*. M. Mignault studied cello at the illustrious Paris

Conservatory, founded in 1795. My music-theory teacher, Josèphe Colle, studied there, too, under the tutelage of the great French composer Olivier Messiaen.

The Quebec conservatory didn't just emulate the grand institutions of Europe—it hired teachers formed in them. No wonder their standards were so high.

My older sister still flinches when I mention our conservatory: "It wasn't a lesson if no one cried." Neither of us guessed why this place hired such impressive performers instead of teachers who knew how to motivate children through joy, not shame. The answer, I discovered, has archaic roots.

The word "conservatory" comes from the orphanages of Renaissance Italy, where foundlings known as conservati (the saved) received music training at state expense. In 1535, the first conservatorio opened its doors, in Naples. But it was the Parisian model that set the tone for the great conservatories of Europe and, later, North America. Founded during the French Revolution, the Conservatoire national de musique et d'art dramatique offered free professional training with strings attached: musicians were obliged to perform at state functions that glorified the new Republic.

This catch sounded familiar. At my conservatory, tuition was also free (if you passed muster with the admissions committee). Here's what I didn't know: It was established in 1942 by provincial leader Adélard Godbout, a forerunner for the Quebec separatist movement. Like its Parisian counterpart, my conservatory had an agenda to churn out top musicians to make the fledgling state look good.

The draconian policies I had endured finally made sense, from the juried performance exams, regardless of age, to compulsory lessons in sight-singing, music theory, and chamber orchestra. Any mark under 70 percent was grounds for immediate expulsion. Grades were posted for all to see—no doubt to sharpen our competitive edge.

The Quebec conservatory billed itself as the only government school of its kind in North America, boasting a unique pedagogical approach inspired by the classic master-student model (with its unquestioning obedience).

DESPITE MY OPPRESSIVE early training, I'd be the last to disparage those who dedicate themselves to Mozart and Bach. The music is exquisite, and I've met many young performers whose ardent playing transported audiences.

After leaving the Quebec conservatory, I joined a youth orchestra in Ottawa, the National Capital String Academy. The rehearsals were less stuffy and I got to know my orchestra mates. Our principal violinist, Aruna, was studious and wildly talented. Although her technique was impeccable, I never thought about the mechanics of bow strokes or hand positions when she played. Listening to the silky tones she pulled from her violin, I got the feeling she was enraptured by the music and wanted to sweep everyone else away, too. Music took up too much time for us to become close friends, but we always smiled at each other across the music stands. And I had a soft spot for her little sister Rupa, who followed Aruna like a duckling. The youngest in the group, just eleven years old, Rupa displayed a talent on the violin that some described as "almost Mozartian."

Their parents, Anant and Bhawani Anantaraman, came to every concert, every rehearsal. From what I could tell, their lives revolved around music out of sheer joy. Often, I wished my family was more like theirs.

My stepfather spent his free time volunteering for community groups while Mom served on the board of an artist-run gallery. My older sister had quit the violin by then and my hyperactive younger brother showed no interest in learning an instrument (he'd rather blast his cassette tape of the heavy-metal band Twisted Sister). One Saturday when I was fifteen, I played a small solo part in a Corelli *Concerto Grosso* at the National Gallery of Canada, but I didn't search for my family's faces in the audience. I knew that no one other than my mother would come.

That's the thing about classical music. Except in families like Aruna and Rupa's, many kids end up learning a tradition with little connection to the rest of their lives. Parents love sharing videos of their child in a piano recital performing "Für Elise," but back at home, they're boogying in the kitchen to hip-hop or cranking up Arcade Fire. The sports equivalent would be to send a child to fencing lessons while everyone else watches the hockey game.

Few parents realize that enrolling small children in standard piano or violin lessons is like teaching them to read starting with Shakespeare. We don't expect kids to know about grammar and spelling before they can talk, but from the very first music lessons, children are expected to learn the smallest units of music: notes, beats, rests. Ethan Hein, a music-education specialist at New York University, believes this approach starts at the wrong level of abstraction. Taken out of context, notes and rests don't register as music. A more intuitive approach, he said, would be to begin with simple musical phrases from Beethoven, children's songs, or top-forty hits. His approach reminds me of how people pick up instruments in Brazil—by ear, with musical rewards in the moment, not a distant future.

While scoping out public schools for his son, Hein toured a Brooklyn school that prided itself on offering classes taught by musicians from the New York Philharmonic. He watched as a young woman taught a waltz with an oom-pah-pah beat, a rhythm that reached "peak cultural salience" at least 150 years ago. The students, mostly Black and Latino, listened "politely but blankly." The grooves these children knew from home and even from the schoolyard, said Hein, had infinitely more rhythmic sophistication than anything they heard in class that day.

(Teachers may have little say in the curriculum—and their ethnicity may not reflect the musical genres they know and love—but music education does have a diversity problem. In a 2017 survey, more than 90 percent of elementary and high school music teachers in the U.S. were white.)

Standard music education tends to focus on an extremely narrow type of music, said Hein, namely, "the aesthetic preferences of Western European aristocrats of the eighteenth and nineteenth centuries." The system that gave us musical absolutes, such as metronome rhythm and "perfect pitch," is presented as the best and most universally valid. But, in truth, this music reflects the sensibilities of light-skinned people from a certain time and place. It is an "ethnic" tradition, just as much as Indian sitar music or African drumming.

Hip-hop is by far the most popular music genre in America, Hein continued, yet it is "vanishingly unusual" for it to be played in a typical classroom. And even if school music teachers wanted to cover it, few have the knowledge to convey hip-hop as a values system and not just a music genre. Adam Kruse, an associate professor of music education at the University of Illinois at Urbana-Champaign, argues that hip-hop has a lot to offer, encouraging us to "keep it real," "flip the script," and "make some noise"—create our own music instead of passively consuming it.

Hein has done his part by cowriting *Electronic Music School: A Contemporary Approach to Teaching Musical Creativity*, aimed at inspiring kids to make music using digital tools. He believes classical music could become more relevant in today's multicultural classrooms if teachers approached it the way sampling producers do, "as raw material for new expression." Hein points to a sampling of Stravinsky's "The Firebird" (from 1910) in the 1982 rap hit "Planet Rock," a collaboration by Afrika Bambaataa and the Soulsonic Force. Learning about classical music doesn't have to be dreary.

Meanwhile at home, "bedroom producers" of all ages spend hours tinkering with sound effects on their computerized MIDI systems. I knew little about this scene, so I reached out to Patrick Lewis Wilkie, a twenty-five-year-old engineer who makes electronic music on the side and listens to DIY albums for free on YouTube. Instead of shelling out for cover charges at nightclubs, he and his friends let loose in joyous house parties, laying down their own electronic beats on the fly. And, he said with a grin, "No one cares if it would sound good in the morning."

What about the physicality of old-school instruments? Patrick, who studied classical piano as a child, acknowledges that making music on a computer puts "a wall" between him and any instrument he assigns to the interface: "It's like its vocal cords have been cut." He added that having an ear for rhythm, timbre, and pitch matters in electronic music, just as in acoustic music. Then he paused, and I got the sense he was weighing his own thoughts. "When I look at the [electronic] artists I like, they're all the people who had the early music training."

I thought of Billie Eilish, the California teen whose 2019 hit "Bad Guy" topped the charts. Although the media painted her as a natural-born talent, Eilish started singing in the Los Angeles Children's Chorus at the age of eight. She spent more than a year writing "Bad Guy," with help from her older brother in the music industry, and above her piano she has a neon sign that spells "10,000 hours." Hardly an overnight wonder.

Her story made me wonder where future Eilishes will come from. Today's kids are more likely to log hours on an iPad than a piano keyboard or electric guitar. And in a national survey of U.S. high schools, just a quarter of graduating students had enrolled in a single music course during any of their four years in high school. This reluctance seems at odds with the central role of music in the inner lives of teens. But if music teachers fail to offer them what they want and need from music, said Hein, "should we blame the kids for voting with their feet?"

A FRIEND IN MIDLIFE waxed nostalgic. "Remember how we used to listen to the same albums over and over again? I'd sit in my room playing David Bowie, The Clash. I knew every word by heart, every note." She took a sip of wine, and then asked me why people don't listen like that anymore. I didn't have a good answer.

We still love music. If anything, we listen more than ever before—in America, on average, roughly twenty-seven hours a week. Still, technology has drastically changed how we interact with the songs we love. Music is seldom the main event. Instead, says Levitin, "it's become more of a background thing, kind of like auditory wallpaper."

Our phones double as portable deejays tailored to our hectic lives. Scrolling through playlists, we click on the "recommended tracks" that machine-learning algorithms have served up for us. Arguably, streamed music encourages us to listen far and wide. But in this perpetual song rotation, our brainwaves hardly get a chance to settle into a groove. Music has become a frantic thing. Disembodied. Outsourced.

Neil Young ("Heart of Gold") believes technology has squeezed music of its juice. Nearly 85 percent of the recordings we now consume come from streaming services such as Spotify. Streamed audio files such as AAC, Ogg Vorbis, and the older MP3 do not preserve all the audio information from the original recordings. Depending on the quality of our earbuds, we might not notice what's missing, but Young considers MP3s the Wonder Bread of music, stripped of the upper and lower frequencies that give songs their richness and depth. "People can't get what they need from listening to music anymore," he said, "so it is dying."

I wouldn't go that far. Much of the health research on music has been done using MP3s. And in 2021, Apple Music and Spotify announced "lossless" streaming to preserve more of the original frequencies. Tidal, too, offers high-fidelity sound "like you've never heard it before" (unless you grew up with vinyl or CDs). But uptake of lossless streaming has been slow, and while vinyl records have hipster cachet, in 2020 they accounted for just 5 percent of U.S. music revenues.

Live music has morphed as well. At most concert venues, you can hardly see the stage through all the smartphones snapping away. Fans still flock to music festivals, but they go for the fashion scene, VIP pool parties, and ice-cream tacos as much as the live acts. As Ross Gerber, a California-based blues-rocker bemoaned, "The music is almost incidental to the experience."

What a contrast from the music-making I saw in Brazil, where people didn't hesitate to sing or play a set of cowbells when the spirit moved them. Not everyone has lost the all-together-now ethos that our musical species once shared. In growing swaths around the world, though, music has become a product created by specialists and marketed to the rest of us. Capacities for singing and dancing, evolved over millennia, are getting rusty from lack of use. Mine got rusty, too.

IN HER BOOK *The Artist's Way*, Julia Cameron describes the plight of the "shadow" artist, writer, or musician who lives vicariously through others. Some end up as gallery owners who secretly long to paint. Others become music critics instead of singing their own songs.

To me, this shadow phenomenon helps explain the enormous appetite for TV talent shows that build performers up, only to tear them down. Fans of *American Idol* and *The Voice* project themselves onto their favorite contestants, swelling with pride as braver souls give it their all. Talk about a vicarious pleasure. But, as Cameron points out, shadow artists can get nasty towards those with the guts to put themselves out there. *Idol* viewers will abandon their vocal champions at the first off-note and then revel in the carnage. When Simon Cowell, the infamous talent-show judge, smites a quivering contestant with a line like "You've just invented a new form of torture," ratings go up. The cruelty is part of the point.

Hurling an insult like this at an Olympic gymnast or pole vaulter would violate the laws of sportsmanship. But in the blood sport of singing, Western societies delight in separating the gifted from the "talentless" masses. No wonder people shy away from making music.

I have a theory that *Guitar Hero* video games and tuneful TV shows—*Nashville, Treme, Vinyl, Smash, The Get Down, Mozart in the Jungle, Glee*—act as stand-ins for the musical lives we wish we had. But how many people will ever act on these desires? In a quote often attributed to the American poet and physician Oliver Wendell Holmes, "Many people die with their music still in them."

By my early thirties, I no longer tried to find the music inside. I had become a shadow musician, listening instead of playing, watching others instead of picking up an instrument myself.

Whenever I had the cash or free tickets through my magazine job, I caught shows by Spearhead, Red Hot Chili Peppers, the Pixies, Franz Ferdinand, Ben Harper, The English Beat, David Byrne, Buddy Guy. Live concerts put me in touch with the physicality of music, if only at arm's length. I'd snake my way to the front, craving the sight of nimble fingers dancing on strings. (Only later would I learn about the brain's "mirror neurons," which fire when we see the movements of another as if we are performing the actions ourselves.)

I got my fix up to three times a week at jazz bars, dance clubs, and Latin-music cafes. On rare nights without plans, I'd throw a sawdust log into the fireplace of my one-room apartment and play the same CDs over and over again. Sitting alone in front of the fire, I'd gaze into the flames, doing my best to convince myself that maybe just listening could be enough.

— 4 —
Mood Music

ONE EVENING at an art gallery, I bumped into a journalist from Canada's newspaper of record, *The Globe and Mail*. We knew each other from media events around town, and that night she gave me a tip: the Toronto-based paper was looking for an assigning editor in Vancouver. "You would be perfect for this job."

I was hired within days. Still in my early thirties, I could hardly believe my luck—a decently paying job that brought me closer to the arts. (Like a textbook shadow musician, I never strayed far.) Each week I pulled together an entertainment section covering music, visual arts, architecture, and theater. But it didn't last.

The Globe scrapped the section after two years, reassigning me to a role as a health reporter. To my surprise, the new position suited me. I liked talking to scientists, writing features, and staying on top of breaking health news. One morning, a jaw-dropping report about music caught my eye: "Music releases mood-enhancing chemical in the brain." For the first time, brain imaging had uncovered the mechanism that explained why music might have an antidepressant effect. Huh.

My classical training had left me anxious and depressed. The opposite effect. I wracked my brain thinking about how the new findings might have been true for me back then.

A devastating memory from my teen years came to mind, one I never talked about. Flashing back to the terrible event and its long aftermath, I realized for the first time that it was music that had helped me get through it.

THE SUMMER I turned fifteen, I loaded my cello onto a bus for a day's journey from my hometown of Ottawa to a music academy in Charlevoix, Quebec. As the bus lumbered along the St. Lawrence River, I gazed out the window, willing it to go faster. I couldn't wait to catch up with musicians from my youth orchestra, and maybe get pointers from an international soloist or two.

After the sweltering ride, I lugged my cello past a cluster of weathered buildings to the girls' dormitory. I saw many familiar faces as I unpacked my things, but the bunk beds were filling up fast, with no sign of our principal violinist, Aruna, or her whiz-kid sister Rupa from the second violins.

Strange. Normally they'd be here. I called out, "Where are Rupa and Aruna?"

The room froze. Everyone stared.

"Didn't you hear?" one of the girls said at last. "They died in the Air India crash."

What? The room of faces blurred before my eyes. Dazed, I shook my head. "No way," I said. The terrorist bombing was all over the news. But Air India Flight 182 had departed from Montreal—not Ottawa. Rupa and Aruna couldn't have been on board.

If only that were true.

The two sisters had traveled to Montreal to catch the flight. On June 23, 1985, the bomb exploded while the passenger jet was in midair, sending the bodies of all 329 people—Aruna, Rupa, and their mother, Bhawani, too—hurtling into the Irish

Sea. A cellist from my orchestra filled me in. "One of their violin cases was found floating in the wreck."

Oh God. The girl's mouth kept moving but I could no longer make out the words. She sounded muffled, like someone talking underwater. I felt like throwing up. My head throbbed as I swallowed, trying to choke back the sour taste. The two sisters were gone and I hadn't even known.

Everyone must have been talking about it for weeks, but I had kept myself scarce all summer, staying clear of my mom as she prepared to give birth after an unexpected pregnancy. I'd been working full-time for a family friend, running errands and stocking boxes in her clothing warehouse. Days before I left for music camp, my little sister was born in my parents' bedroom.

I didn't know how to explain my ignorance of the tragedy to the girls in my orchestra. While everyone else filed out of the dormitory for dinner, I just stood there, stunned.

No more Aruna? She was the one who played the long, clear A that got everyone in tune, and cued in the first violins with a bob of her dark curly hair. At our spring gala concert, she'd performed Bartók's *Romanian Dances* and a Bach aria—more solos than anyone else. And little Rupa? She was practically our mascot, a beacon in the red jacket she wore.

Music, I discovered, was the inspiration behind the flight. Their mother was taking them to India to give her family a chance to hear her talented daughters play. Their father, a scientist in the Department of National Defence, stayed in Ottawa for work. I could hardly imagine his horror. He had lost his entire family in a flash of light.

Months went by.

The following spring, I performed in the Ottawa Music Festival, my first competition, and despite my nerves, the judges declared me the winner of the trophy for strings. But where was the prize? "Wait here," one of the festival organizers said. I waited so long that volunteers in the auditorium began to stack up the chairs. A door opened and a man walked up to me holding a golden cup the size of a punch bowl. His eyes were sunken and his hair was pure white. Startled, I recognized him. Aruna's father. The last time I'd seen him, his hair was black.

I stared at the hunk of gilded metal in his arms. Aruna must have won the trophy the previous year. I didn't know what to say. He handed it to me, not saying much either. Then he gave me something else: a photo of Aruna holding the golden cup, radiant and beautiful, two months before she fell from the sky.

I thanked him, my face burning.

The Air India bombing is the biggest mass murder in Canada's history. Yet despite knowing two of the victims, I don't remember talking about it with anyone. Back then, no one offered grief counseling to teens unless they showed obvious distress. Normally, I might have confided in my older sister but she had left for university, leaving me with a new baby in the house and my nine-year-old brother bouncing off the walls. Mom would have listened, but she had her hands full. I don't think she realized how much time I spent barricaded in my room, lonely and depressed. To this day, neither my mom nor my older sister remembers my ever mentioning Air India.

My low mood didn't register for me either. The drive to succeed as a cellist consumed me, numbing everything else. I spent most of my time either practicing in my room or mapping out my plan to finish high school in three years instead of four. But whenever I had the house to myself, I'd lie on the couch listening to my idol, Jacqueline du Pré, playing Elgar's *Cello Concerto in E Minor.*

(It would have blown my teenage mind to know I was listening to the same Stradivari cello that Yo-Yo Ma would invite me to play the following year, passed down to him by Jacqueline du Pré's estate.)

This 1965 recording, passionate and brutal, became the soundtrack for my last two years at home. Sir Edward Elgar wrote his cello concerto at a cottage in Sussex near the end of the First World War, harrowed by the sounds of artillery fire coming across the English Channel from France. With this haunting concerto, at age seventeen, du Pré launched herself to international stardom. When multiple sclerosis attacked in her late twenties, it became the anthem of her heart-wrenching life. Over and over, I'd listen and weep.

LOOKING BACK, I'm not sure why I never connected my Elgar ritual to the loss of Aruna and Rupa. All I knew was it made me feel better. Now I understand why: People struggling with depression often gravitate to melancholy music. And it turns out we have healthy reasons for choosing sad songs when we're down.

Psychologists used to consider this behavior maladaptive, a form of wallowing. But Jonathan Rottenberg, director of the mood and emotion laboratory at the University of South Florida, didn't believe people would choose music that compounded their depression. "Their mood state is extremely unpleasant," he pointed out. "They go [into] therapy and say, 'I want to snap out of this.'"

He and a graduate student, Sunkyung Yoon, tested this hunch in a 2020 study using music rated by Western audiences as neutral, happy, or sad. Tracks ranged from Jacques Offenbach's peppy "Infernal Galop" to Samuel Barber's doleful "Adagio for Strings." Overall, people with clinical depression showed a strong preference for somber music, saying it made them feel calmed, soothed, and "even uplifted."

This won't surprise anyone who has found comfort in Mozart's *Requiem* or Lady Gaga's "I'll Never Love Again." Sad songs never pressure us to snap out of it. In a survey of adult listeners, one described how downer tunes helped her "cry a little and then feel relieved, and move on." Another said she felt "befriended" by the music.

Like an empathic friend, sad songs meet us where we're at. And when we're in a funk, chirpy lyrics can feel like annoying platitudes. How many people in a blue mood walk around singing "Don't Worry, Be Happy"? To paraphrase an Internet meme about calming down, never in the history of "Cheer up!" has anyone cheered up by being told to cheer up.

Certain songs always make me misty-eyed, like the late singer Eva Cassidy's soaring cover of Sting's "Fields of Gold." Another is Tracy Chapman's "The Promise." When she sings the words "I'll find my way back to you," she gets me every time.

Poignant music invites us to savor emotions that can be painful but also intensely beautiful. In fact, sad songs may stimulate our body's pleasure responses, including "goosebumps" and "chills," as much as or more than happy music. And as noted in Japanese research, when songs make us weep or put a lump in our throat, music can trigger a cathartic release.

Even in moments of extreme suffering, music offers solace. A journalist friend, Jennifer Van Evra, told me about visiting her neighbor Roy in palliative care as he lay dying of bone cancer. Doses of morphine made him drift out of consciousness, but at times he'd revive in agonizing pain from the disease. On one visit, "he was inconsolable." Knowing he loved music and had once been a churchgoer, Jennifer pulled out her phone and asked if he knew any hymns, "but he was too out of it to name any." She played "Amazing Grace" followed by "Silent Night," and to her amazement, he began to sing along. "You could just see this calm wash over him."

Music reaches us at a level beyond conscious thought. More than any other artform, it is both "completely abstract and profoundly emotional," wrote the late neurologist Oliver Sacks. "Music can pierce the heart directly; it needs no mediation."

EARLY PSYCHIATRISTS had an intuitive grasp of music's mood-enhancing effects. In the mid-nineteenth century, specialists in Baden, Germany, believed individuals with mental illness healed best in a pastoral setting with plenty of music—especially Mendelssohn. The local asylum, Illenau, urged patients to sing in the choir, join the in-house band, and try writing their own compositions. Music and singing, Illenau officials wrote, were "indispensable" to patients as "therapeutic instruments." Months after her stay, one former patient sent a letter requesting copies of her three favorite Illenau songs, saying they reminded her of her restored health and of "feeling all singing-like."

Through the ages, though, no one knew what was happening in the brain when music calmed an agitated patient or roused a listless soldier from a catatonic state. At last, near the turn of the twenty-first century, a critical mass of scientists began to shed light on the mind-boggling chemicals and electrical patterns activated by music.

One of these scientists grew up in Argentina listening to pop music and tango like everyone else. But at age thirteen, Robert Zatorre got his hands on a vinyl recording of music by the Hungarian composer Béla Bartók. He started playing it out of curiosity—and was blown away. "I had chills down my spine. I had goose bumps," he said. "I just felt this unbelievable sensation that I really couldn't explain." That day, he decided he would learn to play music and study it scientifically, too.

Zatorre trained as an organist while earning degrees in experimental psychology. When he joined the Montreal Neurological Institute (the Neuro) in 1981, it was one of a handful of centers in the world doing brain imaging in humans. Zatorre, still at the Neuro, explained to me how music gives us joy, and even euphoria, through some of the same pathways stimulated by chocolate, cocaine, and sex.

ZATORRE AND VALERIE SALIMPOOR, a McGill graduate student, became the first to prove that music triggers dopamine in the brain. Dopamine is the main driver behind addictive behaviors such as gambling, compulsive shopping, and recreational drug use. Dubbed the "Kim Kardashian of molecules" by a British psychologist, this racy chemical prods us to get more of what we want and crave.

When music builds to a peak moment during, say, a drawn-out drumroll, we get a surge of dopamine. Then, if the climax exceeds our expectations—with, perhaps, a spectacular crash of cymbals—dopamine spikes again.

Dopamine isn't the only chemical involved in musical pleasure, though. The brain makes its own versions of heroin, morphine, and cocaine. Known as "endogenous opioids" ("endogenous" meaning "of internal origin"), these chemicals give us everything from a "natural high" to a mild tranquilizing effect. Endorphins, for example, are short for endogenous morphines.

Whether extracted from poppies or made in the brain, opioid molecules behave in similar ways: They attach to tiny receptors throughout the brain and other organs, including the stomach, nervous system, and lungs. Plugged into our opioid receptors, these molecules can trigger a whole-body response to music, like the wave of euphoria fans have at rock concerts.

Along with his colleagues in Spain and France, Zatorre theorizes that music gives us two kinds of delight: intellectual enjoyment and physical pleasure—goosebumps, chills, prickles down the spine. In one study, listeners given a dopamine-enhancing drug said they liked the music significantly more than when they took a dopamine-blocking drug. Dopamine changed their reported enjoyment.

Next, in a prepublished study, Zatorre and colleagues repeated the experiment with an opioid-enhancing drug. This time, music listeners showed strong physical pleasure—goosebumps and chills—yet the drug had little effect on how much they said they enjoyed the music. An opioid-blocking drug didn't change their reported enjoyment much either. Clearly, their bodies responded to music differently than their minds.

While the roles of dopamine and natural opioids remain "very much under debate," said Zatorre, he believes dopamine may be responsible for our mental or aesthetic enjoyment of music, while opioids enhance physical pleasure in music.

This theory makes sense considering how the brain's pleasure-and-reward pathways evolved. Early on, physical pleasures, from sweet foods to sex, helped keep us alive. As the human brain developed, though, we learned to find pleasure in things that required higher-level thinking, such as basking in Brahms. Cocaine and sex give us a rush of pleasure, but we also get hits of bliss from what neuroscientists call "aesthetic" or "cognitive rewards."

Pleasure is life-affirming. In contrast, a lack of pleasure in normally enjoyable things is a hallmark of clinical depression. But tinkering with the brain's pleasure chemicals in a lab isn't enough to prove that music can lift depression or soothe anxiety. For this, we need documented mood changes in real people. Fortunately, we do have studies like these. Loads of them.

OVER COFFEE with a new friend, I asked what she'd want to learn about in a chapter I was writing on music and mood. "Anxiety," she said, "because I have it."

Despite holding a driver's license for nearly two decades, up until four years ago, Liliana Moreno seldom got behind the wheel. As a child in Colombia, she was riding in the back of the family car when her father rounded a sharp corner—and collided with a bus. No one was injured, but the accident made her so skittish that she avoided driving until her son outgrew the after-school programs they could reach by bus. Luckily, she found something to soothe her nerves: music. Whenever she puts the key into the ignition, she plays chill tunes from artists such as Rüfüs Du Sol, an Australian group, or Nora En Pure, a deep-house producer born in South Africa. "It helps," she said. "It's my therapy."

When I mentioned that her remedy has a scientific basis, she beamed.

The evidence comes from surgical wards, where patients with acute anxiety end up with more pain, a higher risk of infection, and longer recovery times. Although sedatives calm most patients, they also carry the risk of breathing problems, blurred vision, dizziness, and agitation. Anesthesiologists searched for alternatives.

At a Barcelona hospital, one group of surgical patients received a standard dose of Valium. A second group listened to half an hour of classical or new-age music, both the day of the procedure and the night before. Just before the surgeries, researchers measured patients' blood pressure, heart rate, cortisol, and anxiety levels. They found no difference between the two groups. As a treatment for preoperative anxiety, they concluded, music was "as effective as sedatives."

A lone study, however, shouldn't convince anyone to swap Valium for Norah Jones. That's where Cochrane comes in. This global network of evidence-based research conducts stringent reviews of dozens of studies to weed out dodgy health information. In four separate reviews of music for preoperative anxiety—the most recent covering twenty-six studies—Cochrane confirmed that music offers a "viable alternative" to standard sedatives.

Music may not soothe every soul, as some may be less responsive to its calming effects. Those suffering from severe anxiety, from phobias to post-traumatic stress, should seek professional help. Still, if music can compete with tranquilizers in a nerve-wracking hospital environment, in my eyes, it's potent enough to take the edge off garden-variety anxieties, such as preflight jitters.

Then there's stress. We tend to lump anxiety and stress together because both cause sleepless nights, fuzzy thinking, headaches, and irritability. Anxiety encompasses everything from acute fears to persistent phobias. Stress, on the other hand, starts as a physiological response. When we're under threat, cortisol raises blood sugar levels for quick energy, while adrenaline causes our heart rate to quicken, readying us for "fight, flight, or freeze." If the threat persists, our bodies stay on high alert, keyed-up in a state of chronic stress.

Here, too, music can dial us down. A Dutch review of 104 clinical trials described music's "moderate tranquilising" effects as "very significant" for preventing and treating symptoms of stress. It didn't matter whether people worked with a music therapist, heard live music in a group, or listened to recorded music alone. Based on results in a total of 9,617 participants, music lowered heart rate, blood pressure, and cortisol levels, along with nervousness, restlessness, and feelings of worry.

After a rough day at the office, though, how to choose the ultimate chill tune?

Music at sixty to eighty beats per minute, the pace of a resting heartbeat—a rhythm we hear in the womb—appears to lower stress best. Just twenty to thirty minutes of slow-paced music, noted the Dutch review, has "a direct stress-reducing effect."

On YouTube and Spotify, playlists arranged by beats per minute are easy to find: slow tracks range from Otis Redding's version of "My Girl" to "Take Five" by the Dave Brubeck Quartet. That said, dopamine increases most when we enjoy the music. And a listener's preferences—not the music genre—has the greatest impact on brain connectivity in our default mode network, involved in empathy and self-awareness.

What if we loathe the tunes? Hypothetically, if we find them irritating (like the new-age Muzak I can't stand) even so-called "relaxing" music could ramp up stress (see chapter 7).

One listener's medicine is another's poison.

AS A CHILD, I learned a piece that sprinted up and down the cello in runaway triplets, exhilarating to play. Maybe it was the odd mix of a jolly tune in a foreboding minor key, but I never tired of this "Tarantella." Not until decades later did I learn that the composer, William Squire, took his inspiration from a mysterious illness that had plagued the Mediterranean for several hundred years. While tending rows of tomatoes and spicy peppers, peasants developed sudden breathing problems, melancholy, and a "sensation of dying." They blamed their symptoms on the venom of the European tarantula spider. Frenetic dancing to tarantella music was the only cure.

At harvest time in southern Italy, this musical antidote was in such demand that fiddlers reportedly wandered the fields like mobile first-aid units, ready to strike up a rousing tarantella at first bite. In Spain's La Mancha region, eighteenth-century physicians treated more than fifty cases of "spider sickness" with everything from bloodletting to viper's grass. In the majority of patients, however, only tarantella music restored the will to live.

But this remedy makes no sense. Vigorous dancing should cause the toxin to spread through the bloodstream faster, worsening symptoms instead of relieving them. Adding to the puzzle, many victims showed no sign of spider bites. What to make of this dubious illness and its musical cure?

Modern scholars have described tarantism as a "mass psychogenic illness" triggered by mental distress and spread to large numbers through "social contagion." Belief in the spider sickness gave depressed peasants a socially acceptable culprit for their miseries under feudalism. More importantly, it gave them an excuse to get out of the field ruts and join in a mood cure that literally put a spring in their step.

I'll bet the peasant remedy actually worked. In a large 2017 review, German researchers noted "highly convincing" evidence that music improves symptoms of depression and quality of life.

The poet Emily Dickinson described depression as "a funeral" in the brain. Interfering with work, school, and social relationships, this mood disorder brings persistent feelings of sadness and low self-worth, along with sleep problems, lack of energy, and, often, thoughts of suicide. While up to two-thirds of clinically diagnosed people may improve with antidepressants and talk therapy, in a Cochrane review, music therapy offered an extra boost compared to standard treatments alone.

Of course, music therapy isn't the same thing as moping around listening to sad cello concertos. Music therapists have extensive university-level training in using music to treat physical, cognitive, and emotional issues. To improve hand-eye coordination, for example, a music therapist might ask a patient to play notes on a xylophone. Depending on the condition, from brain injury to extreme grief, people working with a certified music therapist might show improvements beyond what other treatments can offer. With depression, though, it's unclear whether music therapy relieves symptoms any better than music listening "prescribed" by a doctor or nurse (known as "music medicine"). This puzzled me, because more often than not, a health strategy that involves human connection wins out.

In a study of cancer patients with low mood, music therapy and solitary listening offered similar benefits. Many patients preferred working with a music therapist, saying they liked the feeling of camaraderie and support. But others felt anxious or even hostile when a therapist handed them an instrument or asked them to sing. Left alone with headphones, one patient said, "You can concentrate more on your music, and it's like it relaxes you more."

Music, as Oliver Sacks said, "needs no mediation." Our pleasure-and-reward pathways are easily stimulated by rhythm and song. Moreover, some scholars suggest the brain's endogenous opioid system may also be directly involved in regulating mood. While the details are still being worked out, music is proving to be a fast-acting antidepressant. A 2020 analysis reported a "significant reduction" in depression symptoms from twenty- to forty-minute sessions of either music therapy or music medicine. And shorter treatment periods—twelve sessions or fewer—showed the most benefit.

Just about any music can offer temporary relief. Studies have used everything from European classical to Indian ragas, Irish folk to reggae, and lullabies to rock. The genre doesn't seem to matter, as long as people have a choice. The more we like the music, the better our chances of experiencing a mild, depression-lifting euphoria. Depending on our tastes, the most effective musical antidepressant might be anything from a showstopper from *Hamilton* to the golden oldies Grandma used to sing while making pie.

Researchers have described music as an "emerging treatment option" for mood disorders that "has not yet been explored to its full potential." Yet in many parts of the world, this feature of music is blatantly obvious.

The renowned psychiatrist and trauma specialist Bessel van der Kolk emphasized this point in his trailblazing book *The Body Keeps the Score*: "The capacity of art, music, and dance to circumvent the speechlessness that comes with terror may be one reason they are used as trauma treatments in cultures around the world."

Andrew Solomon, a professor of psychology at Columbia University, made a similar observation in Rwanda. When foreign aid workers offered talk therapy after the genocide, Rwandans found these efforts either offensive or ludicrous. In *The Guardian*, Solomon recalled a Rwandan man telling him the aid workers were "re-traumatising people by dragging them back through their stories." He quoted the man as saying, "There was no music or drumming to get your blood flowing again. There was no sense that everyone had taken the day off, so that the entire community could come together to try to lift you up and bring you back to joy." Instead, Western therapists would take Rwandans one by one into "these dingy little rooms" and have them sit for an hour and talk about the horrific things that had happened to them. "We had to ask them to leave."

(Since then, global organizations have made efforts to provide culturally sensitive programs. In 2012, a group called

Musicians Without Borders launched Rwanda Youth Music, offering drop-in drum circles, therapeutic music sessions, and music camps for youth coping with trauma, poverty, and HIV.)

Keep in mind, though: listening to music alone at home may not offer the same benefits found in medical settings or group traditions grounded in community support. For one thing, music "prescribed" by a health-care worker might be more likely to cause a placebo effect. Michelle Dossett, a researcher in Harvard University's Program in Placebo Studies and the Therapeutic Encounter, confirms that a patient's trust in a practitioner, just like a sugar pill, can be enough to trigger mental and physiological responses that may support their recovery. Left on their own, people with depression often struggle to do the things that make them feel better, such as going for a walk or listening to music. Many need outside support, along with professional help.

And in times of acute trauma, as I learned from one family's story, even the gentlest of songs might be too much to bear.

ON THANKSGIVING DAY in 2019, Leila Viss and her husband, Chuck, were heading home from a church service in Denver when they checked a string of messages on their phone. A Florida police officer was trying to reach them about their twenty-five-year-old son, Carter, who had been hit by a boat. All of his limbs had suffered severe damage. Leila remembers taking a red-eye from Colorado to Florida, crying through the flight. "We didn't know if Carter would live and if they could save his legs."

Their son, a marine biologist, had spent the morning snorkeling with a buddy at a reef. Sun shone through the calm waters as they spotted lionfish, turtles, octopuses, angelfish. Despite the red flags on the surface indicating divers below, a twenty-three-foot pleasure craft roared through the marine sanctuary—straight towards Carter, who couldn't swim away in time. Propeller blades sliced into every limb. His right arm, severed, sank to the seafloor.

By the time Leila got to the hospital, Carter was heavily sedated. He had a tube down his throat, a brace around his right leg, and a bandage the size of a boxing glove around his left hand. Her knees buckled at the sight. Her son's six-foot-three body was missing three-quarters of his right arm. Even as she prayed for his life, Leila worried about how he would cope without two hands to play music to soothe himself.

Carter had just bought a new piano. He'd started lessons at age four, with Leila as his first teacher, and as an adult, he played bass in a church band. Besides his passion for marine creatures and his work at a sea turtle refuge, music was his greatest joy. "When he played," said Leila, "you could tell it was part of him."

It was part of her, too. Leila was a piano professor at Denver University, and a keyboardist in church. Other than her family of three children and her Christian faith, her life gravitated around teaching and performing. After the accident, though, "I could not listen to music." Not even hymns. "The verses fell flat."

Alarmed by her sudden aversion to music, Leila read up on trauma and how it short-circuits the brain. Her mind, she realized, had no room for anything with emotional intensity, imagination, or hope. "I could not let myself go beyond the horror we were in."

At the hospital, Carter caught her off guard when he asked, "How am I going to play piano again?" Leila wanted to break down and weep. "You know," she said, trying to hold herself together, "a lot of people can play fantastic things with one hand."

Carter started searching on YouTube for one-handed piano repertoire.

Meanwhile, Leila nudged herself. She discovered she could tolerate listening to ethereal works by a young Norwegian composer, Ola Gjeilo. Then a friend suggested musical meditations by David Baloche inspired by passages from the Bible. If someone had quoted the same scripture, "it wouldn't have mattered," Leila said. But somehow, Baloche's ambient arrangements "cracked open my soul just a little bit."

Leila was staying in Carter's condo, where his new piano was collecting dust. Six weeks after the accident, she got herself to practice a Brahms rhapsody. Then, "I decided to start being creative at the keys." Notes poured out as she interpreted the accident musically, from the first terrifying nights in the hospital to the weeks of medical unknowns. She named the piece "Angel94" after her son's favorite fish, his birth year, and the passcode they used to get into his hospital room. Expressing her emotions in music helped her get through "a really rough time."

Leila's husband stayed with Carter through the last of his seven surgeries and months of occupational therapy to restore function in his legs and remaining hand. Leila, back in Colorado with their other children, tried to count their blessings: Carter had survived and was regaining his ability to walk. In August, nine months after the accident, Carter sent her a video of himself playing—with one hand—a piano arrangement of the prelude from Bach's *Cello Suite no. 1*. Tears streamed down her face.

When Leila shared this video with me, I watched in awe as Carter's left fingers fluttered over the keys, capturing every note of the cello piece I knew so well. Part of me kept noticing the hand that wasn't there, but mostly, I was struck by his physical grace and conviction at the piano. I imagined the hours he must have spent retraining his hand to play the contrapuntal melodies with such skill and tenderness.

While Carter continues to heal, Leila is writing a book about her family's ordeal in hopes their experiences might help others, because "if you don't have a trauma yet in your life," she said, at some point "you probably will." Besides her faith, Leila singles out music as the most important "intervention" in her own healing. She often mentions the famous Aldous Huxley quote: "After silence that which comes nearest to expressing the inexpressible is music."

DELVING INTO THE NEUROCHEMISTRY of music sharpened my understanding of its mysterious influence on human emotions and mood. At the same time, it made me ask myself, again, how I became so unraveled.

If music lifts anxiety and depression by stimulating pleasure chemicals, then playing the cello day after day should have made me the picture of mental health. Instead, it drove me to burnout and despair. Did I have a glitch in the pleasure pathway in my brain?

A conversation with Barry Bittman, an American neurologist, gave me more insight. Bittman conducted some of the first studies demonstrating that music can strengthen our immune response (see chapter 6). But under the wrong conditions, he explained, music can have the opposite effect.

Bittman had divided non-musicians into three groups: The first joined a drumming circle for half an hour. The second sat in a circle and listened to drumming music, while the third read newspapers and magazines. Initially, Bittman found no differences in their blood samples. "Not a damn thing." He asked the leader of the drum circle to talk less so people could drum more. This time, participants' immune markers "actually worsened."

Bittman was nonplussed. Then he figured out that most of these non-musicians had either had a negative experience with music earlier in life or were convinced they weren't musical. Being asked to drum on the spot was stressing them out: "We were actually pushing their immune systems in the wrong direction."

Bittman scrapped his approach. Next, the beginner drummers pounded out the syllables of their names, played with shakers, and jammed however they liked. "When we removed that sense of fear or anxiety about performing on an instrument," he explained, "biology changed." Participants showed a surge in natural killer cells, the specialized white blood cells that seek and destroy pathogens and cancer cells. Given the freedom to make their own music, he said, they loosened up and embraced a spirit of fun and camaraderie. "That's when the magic happens."

With these words, everything clicked. My music training was almost guaranteed to induce a stress response—enough to dampen my immune system, along with my pleasure pathway. I loved music, loved the sound of the cello. But I almost never played with a sense of creative freedom or fun. Through all those years of persevering, I had missed out on some of music's greatest gifts. This realization hurt.

I had done it all wrong.

5
A Musician's Brain

REACHING INTO A WICKER BASKET, my mother pulled out a clay flute shaped like a pair of breasts. She held it in her hands, admiring it for a moment, and then placed it on the table next to a dozen other flutes she'd made in the Mayan pottery village when I was small. One of these instruments looked like a giant tobacco pipe. As my fingers traced the length of rough unglazed clay, Mom explained that the bowl at one end was for holding water. I went to the sink to fill it, and marveled at the gurgling sounds her creation made.

These flutes had been wrapped in newspaper, squirreled away, until I pressed Mom for details about her spontaneous music phase. After my failed attempts to jam on the cello, I doubted I'd ever learn to improvise, but the recent discoveries about rhythm and song had kindled my curiosity.

As a teenager, Mom recalled, she had considered a career as a pianist or composer but decided she didn't have the talent. The spontaneous music scene she encountered in Vancouver in the early 1970s was "amazing," she said, because "I played my own sounds for the first time." Encouraged by Michael and Johnnie, she could ignore her nine years of formal piano training, and "actually toss it aside." She paused and unwrapped another flute. "It was one of the most important things in my whole life."

Improvising with music catapulted her to a new level of creativity, she said, allowing her to break free of her formal art training as well. I'd grown up admiring her vibrant frescoes, breathtaking landscapes, and glittering mosaics, never knowing that spontaneous music-making had jump-started her creative life. This revelation floored me.

The more I thought about her blissful immersion in flower-power sounds, the more mystified I was by the musical path she'd chosen for me. What happened to the values of spontaneity and free expression when it was my turn to learn music? "Didn't you notice my cello teacher browbeating me each week?"

Mom, put on the spot, questioned whether my training was all that bad. "I mean, my piano teacher used to whack my fingers, you know." I reminded her of my conservatory's juried performance exams for primary schoolers and the constant threat of expulsion. "Yes," she conceded, "the conservatory insisted it was training people to be professional musicians, even if you're five years old. That was a problem. But I didn't have money for another option."

At this point, I had to ask: Wouldn't going without music lessons have been better than subjecting a child to such a rigid system? "I didn't think so," Mom replied. After all, she'd learned to play Bach and Beethoven through formal training, "and I still valued that." Because, she continued, "you learn different combinations of sounds, and how to distinguish them from one other, and rhythms—and on and on," she said. "In my mind, that's huge, valuable mental training."

I looked at her, gobsmacked. It all boiled down to her conviction that music was good for the brain.

Many people share the belief that music makes us smarter. Music is so closely tied to higher thought that Carl Sagan, the astrophysicist, convinced NASA that the *Voyager* spacecraft should carry songs to the stars. Both *Voyager* probes, launched in 1977, contain a gold-plated copper record emblazoned with the words "The Sounds of Earth." Encased in an aluminum sleeve, these phonograph recordings offer greetings in fifty-five languages, the songs of humpback whales, and ninety minutes of music—from Mozart to Chuck Berry to Peruvian panpipes and Senegalese percussion. More than any other sounds, music filled this celestial "message in a bottle" created to give extraterrestrials proof of intelligent life on Earth.

As our songs float around in interstellar space, here on Earth, neuroscientists are still unraveling the mysteries of music and human cognition. Does music truly enhance our intelligence? When we listen to Mozart or play Miles Davis riffs, how much of their genius osmoses to us?

AROUND THE TIME I quit the cello, three dozen college students filed into a lab in Irvine, California, to take part in an unusual experiment. The lead researcher, Frances Rauscher, was a red-haired woman in her late thirties and a former child prodigy. A decade earlier, she had abandoned her career as a concert cellist, burned out by the grind of performing gala recitals in Paris and New York. In her new life as an experimental psychologist, she dedicated herself to studying the cognitive benefits of music—especially Mozart.

For the small pilot study, she selected Mozart's *Sonata for Two Pianos in D Major, K. 448*, a work described by the musicologist Alfred Einstein (a cousin of Albert) as "one of the most profound and mature" of all Mozart's compositions. Students in the lab spent ten minutes either listening to Mozart or a relaxation tape, or sitting in silence. Straight after, they completed puzzles from a standard intelligence test, such as picturing how to cut and fold paper into specific shapes. When Rauscher analyzed their scores, the results took her breath away.

Just ten minutes of Mozart seemed to improve their spatial reasoning by the equivalent of eight to nine IQ points. Although the gains evaporated after ten to fifteen minutes, in 1993, Rauscher described her study in a two-page letter in the journal *Nature*. The media went wild.

Her results—dubbed the "Mozart effect"—inspired a best-selling book and a cottage industry of "brainpower" products, including a slew of Baby Genius CDs. Rauscher, however, never claimed that Mozart increased intelligence. She and her coauthor merely hypothesized that Wolfgang's music "primed" different brain regions for abstract reasoning. But these nuances fell on deaf ears. "You can never control what the marketers will do," she told the *Los Angeles Times* as the Mozart fad took off. "It is a very scary thought."

Mozart mania reached fever pitch in 1998, when Georgia state governor Zell Miller announced a proposal to provide every newborn with a classical-music cassette tape or CD. "No one questions," Miller declared, "that listening to music at a very early age affects the spatial-temporal reasoning that underlies math and engineering, and even chess."

On the contrary, legions of scientists questioned the notion that listening to Mozart could boost IQ. When I looked up the stringent follow-up studies that failed to replicate Rauscher's results, it was clear that the Mozart myth didn't hold up.

But the story of music and brainpower was just getting started.

IN THE THREE DECADES since Rauscher published her tiny Mozart experiment, thousands of people have undergone brain scans and cognitive tests aimed at identifying how music shapes our gray and white matter. Listening to Mozart doesn't raise IQ, but could playing an instrument make us smarter?

In study after study, children and adults with extensive music training tend to outperform non-musicians on tests of working memory, attention, and executive functioning, the higher-level thinking involved in problem-solving and switching between tasks. Learning an instrument at a young age has been linked to stronger auditory processing, emotional perception, and "stick-to-itiveness," all of which may contribute to future success.

But here's the hitch: pegging better life outcomes to a single activity or personality trait is a tall order—and we've seen other valiant attempts end in failure. Take the "marshmallow test," the famous 1972 experiment developed at Stanford University. Children around age five were given the choice to either snarf down one marshmallow straight away or, if they waited fifteen minutes, receive two marshmallows (those with a salt tooth were offered pretzel sticks). In follow-up studies, researchers linked the ability to delay gratification in childhood to better life outcomes, based on SAT scores, educational achievement, body mass index, and other health measures. Years later, however, when the test was repeated in a much larger and more diverse population, researchers realized that children from privileged families had an easier time resisting the treat because they knew another marshmallow would always come their way. Better life outcomes had to do with affluence, not self-control.

With this in mind, does music training alone explain why children who play instruments show stronger reading skills and academic achievement?

I was one of those kids. Starting in kindergarten, I pulled straight As in French immersion, followed by enriched and gifted classes through high school. But how to separate nurture from nature? My father, a non-musician, had the highest grades in mathematics in the Soviet town where he grew up; by age twenty-five, he was a college professor in math. My mother, no slouch herself, worked with scientists as an artist-in-residence on research expeditions in the Arctic and the deep seas.

Then there's the nurture side. If a developmental psychologist had visited my childhood home, they would describe it as a "stimulating environment." (I'd use another word: "chaotic.") Mom raised me and my three siblings in a warren-like abode crammed with paintings, sculptures, engine parts, and curios collected by my stepfather, who had an intractable garage-sale habit. In the basement of our home in central Ottawa, Mom had a hydroponic alfalfa sprout farm—twelve bathtubs full—which supplied local health-food stores. On the main floor, rabbits hopped about, leaving pellets on the tattered Persian carpets in the dining room. (We didn't shit where we ate, but the bunnies did.)

The "idiot box" was mostly off-limits, so I'd pound clay on the dining table, sneak off with my parents' books, from *Narcissus and Goldmund* to *The Joy of Sex*, and teach myself how to sew my own clothes on an ancient Singer sewing machine (another garage-sale find).

Embarrassed by the mess, not to mention my rebellious younger brother, I seldom invited friends to our cluttered quarters. Instead, I met many of my social needs through my parents' entourage of artists, musicians, philosophers, paleontologists, federal bureaucrats, and a depressive poet or two.

With all these early childhood influences, I can't see how a researcher could tease out the effects of classical-music lessons on my report cards. Other children, too, are shaped by a wide array of genetic, socioeconomic, and cultural dynamics. Although researchers do their best to rule them out, correlation studies remain an inexact science.

This much we know: playing an instrument is a brain-twisting feat. Reading music involves translating symbols on the page into specific sounds in the flash of an eye. Performing without sheet music ("learning by heart") recruits muscle memory and several other types of recall. In just two and a half minutes of music, Bach's famous cello prelude in G major has roughly 650 notes with different pitches, bowings, and musical inflections. If a cellist had to consciously remind herself how to play each note (*this one is played with my third finger, on a down bow*), she'd never pull it off.

With cognitive demands like these, it's hard to imagine how playing an instrument wouldn't have an impact on a child's scholastic performance. And yet, solid proof remains elusive.

To prove cause and effect, scientists start with two similar groups: one gets the experimental intervention while the other does not. After a set period of time, researchers compare them and check for unforeseen factors that might explain any differences they find. This tried-and-true method, applied to music training, has yielded inconsistent results. The next step: meta analysis.

In a 2020 study, a team of statisticians analyzed fifty-four studies of music training published from 1986 to 2019. After crunching the numbers, they found no relationship between music lessons and enhanced cognitive skills or academic performance, regardless of the children's age or amount of music training. The statisticians chalked up the benefits found in previous research to a faulty interpretation of the data, along with possible "confirmation bias"—the tendency to find what you're hoping for. The team described efforts to enhance academic skills through music training as "pointless."

Over the long haul, who sticks to music lessons? Kids with personality traits such as conscientiousness and openness to new experience. Pre-existing differences, Canadian researchers have concluded, likely explain the link between music training and higher grades.

While playing an instrument doesn't increase IQ, it can alter the very structure and density of our gray and white matter. As the late neurologist Oliver Sacks wrote: "Anatomists today would be hard put to identify the brain of a visual artist, a writer or a mathematician, but they would recognize the brain of a professional musician without a moment's hesitation."

WHEN NEUROLOGISTS CONDUCTED the first MRI scans of a musician's brain, starting in the early 1990s, they expected to find musical abilities in the brain's right hemisphere and language in the left. To their surprise, musicians showed as much left-side dominance as non-musicians. But the structure of the musicians' brains revealed startling differences: namely, a thicker corpus callosum—the fibrous nerve bundle that sends brain signals back and forth between the two hemispheres—and more gray matter in the auditory, sensory, and motor areas. A musician, declared the neurologist Gottfried Schlaug, "is basically an auditory-motor athlete."

Did these differences come from music training itself, or were those born with a thicker corpus callosum simply more drawn to playing an instrument? This chicken-or-egg question was easily solved.

In a landmark 2009 study, a group of American first-graders attended fifteen months of private piano lessons and practiced regularly at home. A second group had a weekly school music class but no private lessons or home practice. All were closely matched in age, brain structure, and socioeconomic background. After fifteen months, though, only the private piano lesson group showed the same neurological signature as adult musicians. A separate study of adult identical twins found similar brain changes in those who played an instrument but not in their non-musician siblings. The bottom line: a musician's brain comes from nurture, not nature.

Music wires the brain in specific ways, depending on the instrument we play. Pianists have extra gray matter in visual-spatial areas, used to figure out where objects are in space. Drummers have hyperefficient motor areas, allowing them to perform complex movements with far less brain activity than non-musicians. And in classical string players, the left digits, used for fingering the notes, take up extra space on the brain's cortical "map" for processing touch in distinct body parts.

I'd be curious to see brain scans of people who make music using computer software instead of musical instruments. So far, researchers have focused mainly on classical musicians, reflecting, perhaps, an age-old snobbery. But in 2018, neuroscientists scanned the brains of professional beatboxers, who use their voices to mimic percussion sounds. Most of the beatboxers got started at around age fourteen—double the typical age for a professional pianist. Nevertheless, when beatboxers listened to tracks by a fellow performer, Reeps One, the brain area that controls mouth movements lit up. Although they lacked the thicker corpus callosum of an instrumentalist, the beatboxers showed functional brain changes that matched their musical skill set.

Functional changes have to do with how we're using our brain, whereas structural changes refer to the brain's shape and density (more or less white or gray matter). While the fringe benefits of playing an instrument remain unclear, researchers do know what it takes to sculpt the structure of a musician's brain—and it's not just a matter of practice, practice, practice. Timing is everything.

A Montreal study found stark differences between musicians who began training before the age of seven and those who got started between eight and eighteen. Both groups had similar years of experience, but the early learners had greater connectivity between brain hemispheres and stronger ability to time their movements to rhythm. This study identified the years before seven as "the developmental window" for a musician's brain. Other researchers expanded it to around age nine, but regardless of the precise age, neurologists believe a musician's brain is shaped through "synaptic pruning."

Synapses are tiny pockets between neurons that relay electrical and chemical messages, allowing neurons to communicate. When these connections get too little use, the brain "prunes" them to increase efficiency. Constant stimulation strengthens our synapses and neural pathways. This process, called neuroplasticity, occurs throughout life, but never as dramatically as in childhood and adolescence.

Even so, I don't see much point in declaring "last call" for a musician's brain. For one thing, music training may still lead to functional brain changes at a later age (see chapter 9). For another, there's a lot more to music than perfect accuracy or virtuosic feats.

HAVING THE BRAIN of an early learner might help a violinist achieve Paganini speed, but performing music isn't an Olympic sprint, where the athlete either beats the clock or not. The finest musicians aim for artistry. And even in the high-stakes world of classical music, starting in kindergarten isn't the be-all and end-all. At McGill University, I knew a stand-up bass player who started at fifteen—relatively speaking, a geriatric age. Later, while I crashed and burned after graduation, he auditioned and won a coveted spot in the Toronto Symphony Orchestra.

In the story of musical talent, brain imaging gives us a fuzzy outline at best. As compelling as neuroscience may be, I worry about the current emphasis on brain training through music, the earlier, the better.

Anita Collins, an Australian music educator and author of *The Music Advantage: How Music Helps Your Child Develop, Learn, and Thrive*, believes music training plays a vital role in child development. Learning music builds persistence and resilience because "you get it wrong more than you get it right," she says, "and you have to find your own inside motivation to keep it going." In theory, music lessons help children develop tolerance for failure.

While it's true that music molds young minds, I would add a strong caveat: Practice doesn't necessarily make perfect, especially if it means hours of mindless scales, arpeggios, and finger drills. Instead of building tolerance for failure, endless repetition makes children zone out. Who can blame them? Mechanical exercises don't feel or sound musical. And when kids go on autopilot, they end up practicing the same mistakes and stiff hand positions over and over. After doing this for years, I'd argue that ossified approaches to training increase the risk of injury instead of building technique.

I envy the self-taught rock stars whose yen for music carried them through the learning curve. Dave Grohl, drummer for Nirvana and the Foo Fighters, took a single music lesson before deciding to build his own chops. "I taught myself how to play the guitar, I taught myself how to play the drums," he told *Rolling Stone,* "and I kind of fake doing both." David Bowie taught himself to play piano, ukulele, and a stand-up bass made from a tea chest. While self-taught musicians might end up with atrocious technique—Jimi Hendrix's wacky ambidextrous style would make a guitar teacher shudder—at least they never get their passion for music beaten out of them. There's even some evidence to back this up: Self-teaching has been linked to higher motivation to play music. Less restrictive musical environments, noted a 2013 study, "tend to enhance creativity."

Don't get me wrong: I have nothing against music lessons, especially with an enlightened teacher. Over the past century, pedagogues have tried to make classical-music training more intuitive. The Suzuki method teaches kids to learn by ear instead of starting with sheet music, and encourages group playing, often in unison. The Kodály method considers singing the most natural introduction to music and uses clapping, vocal syllables, and hand signals to teach rhythms and do, re, mi. Music for Young Children, a newer approach, teaches musical concepts through games, craft projects, and group playing in piano lessons aimed at covering four different ways of learning: auditory, tactile, analytical, and visual.

Each method has its strengths, but from what I've seen, the pedagogical approach matters less than the warmth of the teacher, the expectations of parents, and the temperament of the child. Here's what gung-ho parents and teachers should keep in mind: Neuroplasticity is neutral. It can reinforce negative habits just as much as positive ones.

My early training not only gave me a musician's brain, but also ingrained a set of damaging thought patterns that have shadowed me since childhood.

ON A BAD DAY, I am rigid and uptight. No matter what's in front of me, from choosing a paint color to throwing a dinner party, I will fixate on getting it right. The trouble is, once I'm gripped by the dread of getting it wrong, I forget to ask myself what I mean by "right." It's as if a switch has been flipped in my brain, short-circuiting the part of me that would normally say "This might work" or "Let's give it a try." Paralyzed by micro-decisions, I can spend hours staring at paint chips or rewriting a letter. Then I'll snap out of it, bewildered and ashamed by the time I've spent agonizing over details that don't matter.

My Little Miss Perfect side stresses me out and keeps me from trying new things. She whispers that I'll never live up to my wildly creative mother, who has always indulged her curiosity and delight in the absurd. One day she left the house with a melted vinyl record on her head, set at a jaunty angle as a kind of Dadaesque fascinator. As a teenager, I was mortified. Years later, I wished I had the guts to not give a damn.

In our bohemian family, with no one steering the ship, classical music gave me a much-needed source of order, structure, and predictability. But did these rigid conventions breed control-freak tendencies? I've often wondered if perfectionism is the natural consequence of immersion in this ultraconservative and technically demanding pursuit. My training left no room for experimentation, body awareness, or exploring the age-old connections between music and dance. The goal was not to enjoy myself or be in flow, but to get it right.

Even as a teenager, I could see how this tradition encouraged a mechanical approach. While attending a masterclass, I took notes on a violinist my age: "He shows not a flicker of emotion and doesn't even twitch when criticized or complimented. He plays very well, but he doesn't put his soul into it. His playing is impersonal, like a robot."

Later, at McGill, I had a summer job as a teaching assistant at a children's Suzuki music camp. I'll never forget a little redheaded boy, nine years old, trembling with his violin. When I tried to encourage him, he burst into tears, unable to receive a single kind word. During the closing performance, the six-year-olds wowed the crowd with their natural musicality. Many of the older kids, though, showed the same stiff posture and grim look I recognized in myself.

None of this rigidity and tradition of fault-finding comes from the music itself. Rather, it's the harsh teaching methods and soulless motivations behind classical training that often trip people up.

I've come to believe that how we approach music matters as much as whether we do it at all. Maybe, instead of giving kids music lessons to make them smarter, we could pay more attention to the links between music, personal expression, creativity, and problem-solving.

As Albert Einstein, a lifelong musician, said in a legendary interview with George Sylvester Viereck in the *Saturday Evening Post*, "Imagination is more important than knowledge."

ONE EVENING IN 1929, Viereck waited outside Einstein's Berlin apartment, and "it seemed to me that I heard strains of elfin music." When the *Post* correspondent entered the mathematician's home, Einstein was wrapping up his violin for the night, "like a mother putting her child to bed."

Einstein declared, with a wistful smile, that if he had not been a physicist, "I would probably be a musician. I often think in music. I live my daydreams in music. I see my life in terms of music." Over glasses of strawberry juice and a shocking number of cigarettes, chain-smoked by Einstein with the "guilty enjoyment of a schoolboy," the writer and the physicist spent the evening discussing the meaning of life.

The mastermind behind $E=mc^2$ insisted that after a certain age, a reading habit "diverts the mind too much from its creative pursuits." But he rarely traveled anywhere without Lina, his trusty violin. Einstein began violin lessons at age six, followed by piano. In his early teens, noted his younger sister, Maria ("Maja"), the budding brainiac would play Mozart on the violin or sit at the piano and "constantly search for new harmonies and transitions of his own invention."

Einstein adored Mozart and Bach, describing their compositions as the epitome of balance and clarity—qualities he strove for in his mathematical work. "Mozart's music is so pure and beautiful," he said, "that I see it as a reflection of the inner beauty of the universe." (These famous words might have helped seed the "Mozart effect" myth.)

Although Einstein himself never claimed that music shaped his scientific work, those closest to him noticed a pattern: "Music helps him when he is thinking about his theories," said Elsa Einstein, his second wife (and first cousin!). "He goes to his study, comes back, strikes a few chords on the piano, jots something down, returns to his study."

Was his musicianship anywhere close to his math skills? To me, this question misses the point. After his 1934 performance in a swanky Fifth Avenue ballroom, a benefit concert for Jewish refugees, *The New York Times* respected Einstein's wish that reviewers refrain from critiquing his playing. He never posed as an exceptional violinist, but his physician, János Plesch, insisted that while he'd known many professional musicians, "there was hardly one whose feeling and understanding for good music was deeper than Einstein's."

Despite his penchant for sailing and mathematics, Einstein told Viereck, "I get most joy in life out of my violin."

PAUL ALLEN, COFOUNDER of Microsoft, played in a rock band called The Underthinkers. In the early days of Microsoft, after hours of writing code, he'd mellow out by strumming chords on his guitar. Music, he said, "reinforces your confidence in the ability to create." Speaking to the writer Joanne Lipman, the late computer mogul and other high achievers in fields ranging from entertainment to information technology credited musical activities for sharpening their capacities to listen, collaborate, and think outside the box.

Nobel Prize winners, too, often moonlight as musicians. After tracking the leisure pursuits of Nobel laureates from 1901 to 2005, researchers at Michigan State University discovered that nearly a quarter of them had either conducted or composed music, or played an instrument.

Long before earning a Nobel for her work on enzyme evolution, Frances Arnold played piano, pipe organ, and guitar. In her 2018 Nobel lecture, she drew parallels between Beethoven's symphonies and the "intricate" and "beautiful" code of life. Richard Feynman, winner of the 1965 Nobel for physics, played the bongos. A bassoon player, Thomas Südhof, was awarded a Nobel in 2013 for his discoveries about tiny sac-like structures called vesicles, a major transport system in human cells. In both music and science, Südhof remarked, it takes rigor and creativity to achieve the extraordinary.

Research on problem-solving emphasizes the value of cross-pollinating our pursuits. When we're stuck in a mental rut, switching to a different activity, like music, can help us approach the conundrum more creatively when we get back on task. (Einstein, as usual, was ahead of the curve.)

Devon Hinton, a psychiatrist at Harvard University, believes music can act as a "flexibility primer," nudging us to perceive ourselves and the world around us in a fresh light. An unexpected key change or a sudden bridge to the chorus can send our minds in new directions. Braided together in music such as jazz, contrasting rhythms and melodies invite us to follow one strand or another, or experience them as a whole. At a sensory level, Hinton explains, we stop focusing on a single "right way" and discover the richness that comes with embracing many possibilities at once.

Learning about "flexibility primers" made me think of all the times I've stared at a computer screen, paralyzed by writer's block. And the time I kept dyeing bolts of fabric to get the perfect shade for a cushion I wanted to sew. What had started as a fun project ended in dark obsession. But never once did it occur to me to break the perfectionism spell by listening to music or picking up the cello. In my case, playing the cello might have made it worse. I'll bet Frances Rauscher, the former cellist behind the Mozart experiment, would understand. Asked by the *Los Angeles Times* why she kept her cello locked in its case, she said she'd given it a whirl a few years before "and it sounded horrible." She could no longer put in the hours she'd need to polish up her skills, and "playing is too much a part of my self-concept to do it badly."

I doubt rusty proficiency was the only thing holding her back. Rauscher mentioned the pressures she'd gone through as a child being groomed for a performance career. "I was wracked by guilt when I wasn't practicing," she said. "Even when I was eating dinner, I felt I was doing something wrong if I wasn't practicing."

A burned-out musician can't easily access the musical outlet that Microsoft's Paul Allen turned to for creative inspiration. For me, this was a cruel irony. I longed to explore the frontiers of my imagination, be a trailblazer like my mother. But my inner control freak kept getting in the way.

In my early thirties, exhausted by this existential tug of war, I went into therapy with a world expert in perfectionism. People with this psychological barrier tend to ruminate over their failings, an unconscious habit that leads to feelings of inferiority, deficiency, and hopelessness—a strong predictor of suicide. Although I never attempted to take my own life, in the blacker moments of my twenties, the thought did cross my mind. If I could never make the grade, what was the point of carrying on?

To my utter disappointment, my new psychologist said there was nothing I could do to tackle perfectionism other than to trust in the therapeutic relationship. Week after week, for five years, I talked about my problems—the failed magazine pitch, the bully at work—and he asked how these setbacks and conflicts made me feel. I had no way of knowing if these sessions were doing any good; they certainly didn't bring me any closer to music. In hindsight, though, I give them credit for opening me up to something else.

While in therapy, I met my older sister at a coffee shop for the first time in six months. She had a husband, a condo, and her first baby bump. "I'm so happy for you!" I gushed, but couldn't look her in the eye. She'd see my eyes burning with envy and shame. I was thirty-two, still renting a modest apartment and meeting lackluster men online. Trying to sound chipper, I told her I was ready to try a new strategy. "Know anyone?"

She did.

Her neighbor's friend Scott, a tall, bright-eyed Albertan, had just returned to Canada after two years of traveling. We gave each other the sniff test at a potluck dinner in a meeting orchestrated by my sister, unbeknownst to him. Over plates of Greek salad and Mexican beans, he told me about his volunteer job for a seed activist in India and the birds he'd spotted during his six months in a Costa Rican bioreserve. Scott had little in common with any guy I'd fallen for in the past. A mechanical engineer, he was outdoorsy, entrepreneurial, and dedicated to preserving the natural world. As we swapped stories, his eyes caught mine, twinkling me.

We met for a hike, and on our second date, he drove us to Vancouver's Jericho Beach with a pair of river kayaks strapped to the roof. (He later confessed he was testing me to make sure I was "waterproof.") After we explored the shoreline afloat, I watched from the beach as he showed me his moves. He thrust his kayak up and down in the water from tip to stern, and then flicked his hips to perform several three-sixty rolls in a row—a paddler's mating dance. Seawater streamed from his muscled arms and suntanned face. I was smitten.

Like me, Scott had a complicated family and a tender heart. When I told him about my brother's close calls in the hospital, my washed-up cello career, and my fear of never creating anything of meaning, he held me close. One night he whispered that he'd love me even if I never wrote another word or played another note. I married him.

Scott didn't sing in tune, let alone play an instrument, but he ticked too many boxes for me to dwell on this area of mismatch. After all, I wasn't making music either. And I couldn't imagine a time when that might change.

— 6 —
More Than Meets the Ear

ONE EVENING, not long after we were married, Scott and I talked in circles for hours. By midnight, we were both exhausted and the quarrel had turned into a full-blown argument. I can no longer remember what the disagreement was about, but even if it was something trivial, the tension was high. I'll never forget the surge of panic I felt when he grabbed his pillow and made for the door. "I can't sleep in the same room as you tonight."

My heart pounded. What if Scott was done with me for good? In the past, my fears of abandonment might have kept me up all night, but this time, I did something I'd never thought of before. Grabbing my phone and a pair of earbuds, I pulled the covers over my head, like a child afraid of the dark. Then I pressed Play and closed my eyes.

A bamboo flute called out. Seconds later, a slide guitar, sounding much like a sitar, teamed up with driving tabla drums. The album streaming through my earbuds was a 1968 classic, admired by George Harrison at the height of America's raga craze. I hadn't listened to *Call of the Valley* since my late twenties, yet the flowing rhythms gave me the same tranquil feeling that had comforted me then. In my blanket cocoon, they reminded me that even if my marriage crumbled, I would be all right.

The next morning, Scott and I vowed to never again tackle tough conversations before bed (a promise we've mostly kept). I took technology for granted. It didn't occur to me that earbuds had given me a way to soothe myself without disturbing my husband on the couch downstairs. With a touch of my phone, I had created a private audio bubble—a kind of listening that didn't exist for any generation before me.

ON A SNOWY DAY in January, when I was around ten, a couple of kids came to school after the holidays wearing a blue-and-silver metallic gadget. Teachers ordered them to keep the device tucked away in class, but at recess, everyone in the schoolyard gathered round. Best friends got to try the foam-covered headphones. The rest of us stood and stared as they mouthed riffs from Billy Joel and Air Supply. Riffs no one else could hear.

The Sony Walkman, launched in 1979, was the first portable music player without an external speaker. Designed to play cassette tapes, and later CDs, this Japanese invention had an asteroid-like impact on how we listen to music.

Before the Walkman, hardly anyone wore headphones other than extreme audio nerds. Stereo headphones, invented in 1958, delivered sound to both ears without cross talk from loudspeakers. Sony took this unique acoustic experience and made it mobile. "Everyone knows what headphones sound like today," wrote the late Sony designer Yasuo Kuroki, "but at the time, you couldn't even imagine it, and then suddenly Beethoven's *Fifth* is hammering between your ears."

The novelist William Gibson, author of *Neuromancer*, bought his first Walkman in the summer of 1981. As he roamed downtown Vancouver listening to Joy Division, he imagined a future where machines delivered data with the same "under-the-skin intimacy" of the new music player. A year later, in his short story "Burning Chrome," Gibson named this virtual world "cyberspace."

"The Sony Walkman has done more to change human perception than any virtual reality gadget," Gibson said in 1993, the digital dark ages. I contacted him to ask if he stood by these words. Yes, he replied in an email: "I do still think it's true."

The Walkman's instant popularity took Sony by surprise. Instead of the expected five thousand sales per month, it sold upwards of fifty thousand units in the first two months. Pocket-sized and entirely private, the Walkman offered an experience that had been previously unavailable at any price: listening to music of one's own choosing, said Gibson, "on demand, while doing anything at all, anywhere."

The trouble is, he added—and I could almost hear the septuagenarian novelist chuckling by email—"it's fundamentally impossible for anyone who grew up with this option to imagine what the world was like without it."

Up until the Walkman, noted the writer Matt Alt in *The New Yorker*, "music was primarily a shared experience." Teenagers blared music from boom boxes and transistor radios. Families spun their favorite tunes on furniture-sized stereo systems in the living room. Then all of a sudden, the Walkman gave us permission to tune out. Photos of the New York subway in the 1980s show rider after rider, young and old, gazing into space with wires dangling from their ears. "It's like a drug," said

Susan Blond, a vice president at CBS Records, in a 1981 interview with the *Washington Post*. "You put the Walkman on and you blot out the rest of the world."

This "drug" has dominated our listening habits ever since, through the iPod (2001), iPhone (2007), and assorted knockoffs. Next came streaming via iTunes, Spotify, Apple Music. North Americans have lapped it up: from 2010 to 2021, streaming went from just 7 percent of U.S. music revenues to 84 percent. For many of us, streaming *is* music.

But have these technologies given us a more intimate relationship to music, or simply a more distracted one?

I'm not sure. Bulky home stereos encouraged us to slow down and really listen, entranced by the rich sound quality of vinyl records, cassette tapes, and CDs. On the other hand, until the heyday of headphones and portable music players, no one could jog to Rihanna or listen to a "pump song" just before a job interview. We're tapping in to music like never before.

Luckily for us, we don't have to choose. While I still cherish my CD collection, new discoveries from music cognition research have sold me on the perks that come with earphones and infinite song selections on the fly.

SCROLLING TWITTER OVER COFFEE one morning, I spotted a post from a friend that stopped me mid-sip: "What is your favourite hype song? The song you play when you really need to get pumped up for something."

Hype song? Not a single tune came to mind. But the hundred replies left me convinced that most people have a power anthem up their sleeve. An entrepreneur shared the "walk-up" playlist she listens to just before giving a keynote speech, while a journalist confessed to naming one of her playlists "You got this."

Most of the picks looked like infectious high-energy tracks, from AC/DC's "Thunderstruck" to Technotronic's "Pump Up the Jam." Some, though, showed a vulnerable side, including "Belle" from Disney's *Beauty and the Beast* and "I Have Confidence" from *The Sound of Music*.

Browsing through the songs made me wonder if I was the only one on the planet who had never turned to music before a major deadline or speaking event. I asked my husband if he'd ever used hype songs. "Of course!" he said, and proceeded to list the Headstones and Offspring tracks he played in his kayaking days just before putting in to a treacherous river. How had I overlooked this strategy?

In his memoir *A Promised Land*, Barack Obama mentions drawing strength from music during his first presidential campaign. "It was rap that got my head in the right place," he wrote, especially Jay-Z's "My 1st Song" and Eminem's "Lose Yourself." Both songs are about putting it all on the line.

Then there's Michael Phelps, the most decorated Olympian of all time, who never hit the pool for a race without listening to music "until the last possible moment," he told *The Guardian*. "It helps me to relax and get into my own little world." At the Rio Games, the swimmer's pre-race tracks glided from Lil Wayne to Nero to vintage Eminem. Phelps won another five gold medals that year, along with one silver.

Even so, hype songs seemed like the auditory equivalent of a lucky charm until I came across evidence that priming ourselves with our favorite tunes really can make us feel and act more powerful.

Costas Karageorghis has rigorously studied pump tunes as a professor of sport psychology at Brunel University London. Loud and upbeat music before a competition, he confirms, can have a small but beneficial influence on athletes' mood and performance.

Music can give non-athletes the edge, too.

Derek Rucker, a professor of marketing at Northwestern University, compared the impact of songs rated high in power, such as Queen's "We Will Rock You," with those rated low in power, including Fatboy Slim's "Because We Can." In a study conducted by Rucker and colleagues, college students who listened to the high-powered playlist opted to go first in a debate nearly twice as often as those assigned low-power tunes. Hype songs had similar effects on other confident behaviors, said Rucker: "It appears that listening to music for three minutes can be enough to—snap!—transform the psyche."

While his claim itself has a whiff of hype, I haven't come across any research that proves him wrong.

For my own hype music, I might choose the Mahotella Queens, a trio of South African women who soared from the depths of apartheid to global stardom. To me, everything from their a cappella harmonies to their booty-shaking moves exudes the feeling "nothing's going to keep me down."

Scott and I saw the Mahotella Queens perform in Vancouver not long after we met. We liked one of their traditional Zulu songs so much that we had it playing over the loudspeaker just after we were pronounced married. Only recently did I look up the lyrics. Lo and behold, the words to "Kukhona Intombi" loosely translate as "The maiden has arrived." How had we known this was a perfect wedding song?

It turns out that most people can pick out a love ballad, dance tune, lullaby, or healing song from a culture unfamiliar to them. A team of scientists at Harvard University gathered songs from eighty-six distinct peoples—from the Ye'Kuana of the Amazon to the Anggor of Melanesia—and played them to 750 listeners in sixty countries. A third of the listeners were from the United States, where healing songs are not part of mainstream culture, said Samuel Mehr, the lead researcher behind the 2018 study, and "yet, people are reliably rating songs that are actually healing songs." His lab confirmed the results in a larger 2019 study with thirty thousand participants in more than a hundred countries.

Humans are better listeners than I thought. No matter where we're from, we receive the message in a dance tune or pump song loud and clear.

I LIKE MY MUSIC LIVE. To me, recorded music is like canned peas: veg from the cupboard still has vitamins, but everyone can tell something's lost in preservation. Nothing beats the feeling of bobbing along to the electric energy of musicians whaling on instruments with everything they've got.

Headphones, I'll admit, seldom appealed to me. When the Walkman first came out, I didn't have the cash for such a lavish gadget, nor did I crave one for my long bus rides to music lessons. Most of the time, I already had music playing in my head. My mind would keep replaying a tricky passage or unusual harmony whether I wanted it to or not.

I never gave my lack of interest in headphones much thought until I saw an article quoting a Canadian teacher and musician, Richard Despard, who pointed out that soundwaves physically alter the molecules in the air. Our bodies register these subtle shifts in the environment, whether our minds are paying attention or not. "A room that is filled with the vibrations of music," he said, "is fundamentally different than a quiet one, as far as your body is concerned."

In a recent *USA Today* interview, Yo-Yo Ma noted that our largest organ is our skin. When music moves molecules through the air, the cellist said, "You feel actually touched. It's that tactile, it's that personal—that intimate." His words resonated with me. Earphones deliver pitches inside my head, but in a music-filled room I am surrounded by rhythm and song, totally immersed in waves of sound.

More and more, though, I've been talking myself into a headphone habit, especially on days when Scott and I are both working from our small open-plan home. I've found other reasons, too. I used to go running or walking without headphones because I liked hearing the birds and paying attention to the rumble of approaching cars. I had never heard that music could take exercise to the next level.

DAYS BEFORE a two-thousand-meter event in Birmingham, England, the Ethiopian track star Haile Gebrselassie convinced organizers to play his favorite song during the race—a hyperactive dance number called "Scatman." When the gun fired, the manic refrain of "Ski-bi dibby dib yo da dub dub" blared through the PA system as he sped to the finish. That day in 1998, Gebrselassie shaved an astonishing one and a half seconds off the previous world record. Later, the runner told the media he had chosen this song specifically to give himself a leg up.

Upbeat tunes help us lift more, run longer, push harder. In his book, *Applying Music in Exercise and Sport*, Costas Karageorghis, the sport psychologist, tells the story of the famous bodybuilder and six-time Mr. Olympia winner Dorian Yates. Early in his career, Yates added a staggering 100 pounds of body mass to his starting weight of 180 pounds. How was this humanly possible? At the gym, Yates lifted as many as eleven tons for an arms-chest workout and forty-one tons for legs, taxing himself to the pummeling tracks of Aerosmith and Guns N' Roses. "To some extent," said Karageorghis, "the heavy metal on Dorian's Walkman was every bit as important as the iron on his barbells."

Karageorghis describes music as a "legal performance-enhancing drug." He doesn't say this lightly. In lab experiments, he has shown that running in time to music helps regulate our stride patterns, reducing the micro-adjustments needed to maintain a steady pace. When we run or cycle in synchrony with music, he explained, our bodies use 6 to 7 percent less oxygen than they need to perform the same feat without moving in sync to a soundtrack. Music has a "metronomic" effect, helping to smooth out the kinks in our movement chain.

A 2020 review of 139 studies yielded other key findings about music in exercise and sport. Music makes exercise seem easier—reducing our perceived exertion by about 10 percent—and more enjoyable, distracting us from the voice inside screaming, "Make it stop!" And good news for people who can't stand the soundtrack at the local gym: fast-paced music improves athletic performance no matter who chooses the tunes.

Songs with 130 to 140 beats per minute offer the most oomph, while music at 120 beats per minute stimulates us to get moving. This pace, about double the resting heart rate of a healthy adult, also happens to be the most common tempo in seventy-four thousand pop songs (analyzed in a separate study). As the exercise review pointed out, "It is with tracks at this precise tempo that deejays routinely lure people onto a dance floor."

Who could forget Olivia Newton-John dressed in shiny magenta tights and singing "Let's get physical"? Workouts have come a long way since '80s aerobics and Jazzercise, but there's still no such thing as a silent spin class, let alone Zumba. In a grueling session of CrossFit, pounding tracks give us the extra push we need to get through another round of burpees.

If workout songs had a popularity contest, Eminem's "Till I Collapse" would win hands down. As of 2021, this song had held first place in Spotify's list of most popular workout songs for five years in a row. The lyrics—on not giving in to weakness, and finding inner strength—amount to a motivational speech.

During a race, though, competitive athletes can't always rely on their go-to songs. Both the Boston and New York marathons have either restricted or "strongly discouraged" the use of portable music players for performance reasons and to reduce the risk of collisions with other runners.

But after a race or workout, slow-paced music can help prompt our body's hemodynamic response, which adjusts blood flow to depleted tissues. Ideally, said Karageorghis, the music should start at around ninety beats per minute "and gradually bring you down towards a state of homeostasis, a resting state, with a tempo of sixty to seventy beats per minute." In general, he added, "you have to think very carefully about the exact function that music is serving and select it accordingly so that it serves you well."

Do the same rules apply to mental exertion? I've met teens who swear by tunes as a study aid, and computer coders who would go on strike if they couldn't listen on the job. Many insist music makes them more productive, priming them for peak performance in the cognitive realm. Are they on to something?

STEPHEN KING PENNED his horror blockbuster *Misery* to blasts of Anthrax and Metallica. Gabriel García Márquez wore out his only two records while writing *One Hundred Years of Solitude*: Debussy's *Préludes* and The Beatles' *A Hard Day's Night*. More recently, Stephanie Land wrote her best-selling memoir *Maid* to a playlist that began with "The Mighty Rio Grande" by the American band This Will Destroy You. Each time the song started playing, she said, "I knew that it was go-time."

I can't imagine writing to a playlist—I'd never finish a paragraph. But I'm not the only one who can't think straight with a soundtrack. In his book *The Organized Mind*, the neuroscientist Daniel Levitin breaks the bad news: Listening to tunes while we're doing other things amounts to multitasking. And multitasking is a myth. What we're really doing, he said, is scattering our attention, zipping back and forth between tasks in a mentally fatiguing and highly inefficient way.

Can't we get better at multitasking? Aren't our brains changing all the time? Not fast enough for such a dramatic change in how our mental resources are allocated. Generally speaking, said Levitin, it takes the brain about twenty thousand years to catch up to our current environment, "in terms of how it's encoded in the genome."

Later in life, we become even more distracted by music. In a 2016 study, compared to adults half their age, those over age sixty showed a significant drop in their ability to match names to the right faces when music was playing in the background.

Fortunately, most people don't have to worry about driving under the influence of music. Volunteers behind the wheel in a driving simulator showed little change in their driving performance with tunes playing, regardless of the tempo or who chose the music. Soft music without lyrics had a calming effect but didn't change how taxing the drivers rated the exercise. Listening while driving, researchers suggest, may be a form of "parallel processing" that "does not threaten" the driving task. However, they added, young drivers often play music too loud to test for ethical reasons—and reckless drivers are unlikely to volunteer for studies. So the jury is still out on whether blasting loud music is harmless on the road.

When tasks are repetitive and cognitively undemanding, such as chopping onions, said Levitin, high-energy music can help us stay alert. The same strategy could be dangerous in a high-stress workplace, like air traffic control, but tunes can take the edge off at break times. Levitin recommends listening to relaxing music for ten or fifteen minutes, with eyes closed, to calm the mind between bouts of focus.

A question, though: Doesn't writing *One Hundred Years of Solitude* or sketching a masterpiece qualify as focused work? Jean-Michel Basquiat painted to Charlie Parker and Dizzy Gillespie. The Baltimore artist Shinique Smith creates vibrant collages and textile works to the tunes of Fiona Apple and Living Color. Could all these artists be wrong?

After watching my mother painting to Franz Schubert and Erik Satie in her studio, I have a hunch that in creative fields, too much focus might interfere with the stream-of-consciousness thinking that inspires artists to do their best work. Of course, this nebulous phenomenon would be tricky for scientists to study, and as far as I know, none has tried.

Still, the next time I have writer's block, I might borrow a page from Pulitzer Prize–winning author Viet Thanh Nguyen. He used to write in silence, but when he got started on his debut novel, *The Sympathizer*, he thought, "Okay, let's try this with some music, but not anything too distracting." He listened repeatedly to the minimalist works of composer Philip Glass. "I wanted to have some of the feel of his music in the rhythm of the prose."

Whether Glass's music enhanced the cadence of Nguyen's writing is tough to prove, but the novelist himself believes it invigorated his work. His music-infused novel won the Andrew Carnegie Medal for Excellence in Fiction, among many other accolades. If I were him, I'd never write in silence again.

LISTENING TO MUSIC alters core physiological systems in our bodies, not just our minds.

Concealed in our eyes, ears, mouth, and nose is a crucial antibody, Immunoglobulin A, our first line of defense against viruses and bacteria. A meticulous review of sixty-three studies describes this antibody as "particularly responsive to music," especially when people enjoy the sounds. Music also reduces levels of interleukin-6, a protein linked to inflammation, infection, and cancer.

The key word is "reduce." Studies of immune markers alone cannot prove that music will protect us from catching a cold or flu. Chances are the immune boost is mild and fleeting, like the lift that comes with laughter or spending time in nature. Hypothetically, though, anything that calms the stress response might strengthen our defense against infections.

This brings us to the "mind-body connection," a term often co-opted by dubious players in the wellness industry. (Give them a tuning fork, and they'll claim to cure arthritis.) At this point, however, even the most orthodox of medical professionals acknowledge that our mental and emotional states can influence how sick we become and how well we bounce back.

When Andrew Levin, a neurologist and amateur trumpet player, first heard of the mind-body connection, "I thought it was new-age woo-woo." But as his knowledge of human physiology and neurology grew, he told the *Pittsburgh Post-Gazette*, "I became increasingly convinced that we actually underestimate how profound this connection is."

The depth of this connection keeps surprising me, too. Music can relieve pain after surgery—even if people have no memory of hearing it.

In a large German study, one group of surgical patients listened to music while under general anesthetic. The second group, also fully sedated, wore earphones with no sound. Neither the patients nor surgeons knew who heard the music, but after the music listeners came to, they reported 25 percent less pain than the non-music group and used significantly less opioid pain medication (such as morphine) in the first twenty-four hours after surgery. The recording, *Trancemusik*, was overdubbed with messages such as "The procedure is going very well," but the power of suggestion doesn't fully explain the results. Similar studies using hypnosis alone have not shown such dramatic pain relief.

Fiona Mattatall, an obstetrician gynecologist in Calgary, Alberta, makes a habit of offering music in the operating theater. She even takes requests: a recent "surgery playlist" included Buddy Holly, Shinedown, songs from *Hamilton*, and The Chicks.

A brain-scanning study at Queen's University referred to music as "the oldest known method for relieving pain." Non-musicians endured painful heat applied to the hand while listening to pleasurable music or suffering in silence. Researchers mapped participants' pain levels in relation to activity in the prefrontal cortex, brain stem, and spinal cord. Music works as a painkiller, they concluded, by evoking a pleasure-reward response that activates the body's descending analgesia system. (Essentially, signals from the brain travel through the spinal cord, instructing the body's endogenous opioid system to suppress pain.)

Music is a mild analgesic compared to opioid painkillers, but it is cheap, free of side effects, and does not interact with medications. In cancer patients, Cochrane researchers reported a "large pain-reducing effect" from either music therapy or passive listening to soothing classical music or jazz. (If only I'd known this when I had tendinitis.)

Music can take the edge off insomnia as well. So far, this affliction hasn't hit me, but sleep problems do tend to worsen with age. I'll keep this prescription handy: in another Cochrane review, listening to slow-tempo classical, new age, or jazz improved restlessness, nighttime waking, and problems falling asleep.

Classical music, especially Bach, is the top choice for drifting into dreamland. In a recent survey, more than 30 percent of adults aged eighteen to seventy-nine preferred classical for bedtime, compared to 11 percent for rock and 7 percent for pop. Once we nod off, though, any music may interfere with rapid eye movement sleep, or REM, also known as "dream sleep." Sleep specialists recommend setting a music player to switch off after about twenty-five minutes.

One night of Enya might not cut it. Insomniacs may need several weeks of nightly music before its sleep-inducing properties kick in, for reasons not fully understood. If adding tunes to a bedtime routine was just a matter of conditioning—training our bodies that it's time to sleep—then audiobooks at bedtime should offer similar results. But they don't. There might remain, noted Australian researchers, "mysterious effects of music that science has yet to address."

ONE OF THE MOST UNNERVING auditory experiences I've ever had could not exist without headphones: binaural beats.

Often marketed as "soundwave therapy," binaural beats deliver two steady tones at slightly different frequencies, one in each ear. Moments later, we'll hear a low beat that sounds like it's coming from inside our head. But this third sound is just an auditory illusion. A phantom pulse.

Here's the working theory: Our brainwaves tend to synchronize, or "entrain," with the sounds we hear. But if one ear picks up a tone at 450 Hertz while the other hears 460 Hertz, the brain will perceive a third frequency—binaural beats—at the points where the two soundwaves bump together. Slightly mismatched frequencies confuse the brain.

Scientists haven't fully studied binaural beats, but that hasn't stopped an overkill of apps from marketing them as "digital drugs" with the power to relieve migraines, sharpen mental focus, and blast away stress. YouTube videos feature teens "getting high" on audio tracks. Even Bayer, maker of Aspirin, promotes binaural beats as "good vibes" for relaxation.

But so far, the evidence of any health benefits is thin. One study poking holes in the hype came from the Montreal-based International Laboratory for Brain, Music and Sound Research. Volunteers wearing head caps with electrodes for monitoring brainwaves using EEG began by listening to binaural beat frequencies. Then they listened to similar frequencies mixed on a computer to create a third beat before the sound reached their ears. In the EEG readings, binaural beats did entrain their brains—but not as much as the premixed beats. And neither had an effect on mood.

Brain-imaging studies of binaural beats haven't yielded convincing results either.

This doesn't surprise me, because the mood lift we get from music comes from our brain's pleasure response to enjoyable sounds (see chapter 4). On their own, binaural beats are about as tasty as cardboard, which explains why they are typically sweetened with nature sounds and new-age music. In studies showing benefits from binaural beats, the music itself might be the special sauce.

Unless stronger evidence materializes, I'll keep thinking of them as quirks of our gray matter—a bit of mischief from the ghost in the machine.

IN THE EARPHONE ERA, the brain-hacking potential of music has spawned an entire industry aimed at packaging the perfect tunes for studying, exercise, relaxation, sleep. Tech start-ups call it "functional music" (as opposed to the dysfunctional kind?). The newish category piggybacks on trends in "functional" medicine and nutrition. (And in my experience investigating health fads, "functional" is often a red flag for pseudoscience.)

A music-meditation app called Meya beckons us to discover the "power of inner alchemy" through a blend of "carefully calibrated" dance music and "neurolinguistic programming," a discredited approach to psychotherapy conceived in 1970s California. Meya's own trial data—neither published in a journal nor peer reviewed—said the tracks helped users feel more "uplifted" and motivated. But the truth is, any music we like can brighten our mood (without the science-y blather).

This doesn't mean all functional music apps are bunk. A venture called the Sync Project matched songs from a streaming service to a Fitbit-like device that tracked biometrics such as heart rate and blood pressure. The gadget signaled to the streaming service when the listener needed to perk themselves up or calm down—with, say, Coldplay for a pick-me-up or Summer Walker to soothe a runaway heartbeat. Using machine learning, the app zeroed in on which musical beats, timbres, or harmonies had the desired effect on the listener's vital signs. The promise: "personalized music therapeutics."

The Sync Project fell off the radar in 2018, when the audio giant Bose snapped it up, but similar products keep coming to market. When I told my biologist sister about them, she said, "Cool! Sign me up." She's the type who likes tracking hours between meals and steps per day, and she'd rather try a machine-learning approach than spend time creating playlists to meet her moment-to-moment needs. I'll bet she's not alone, so I hesitate to knock apps that help draw data-minded types to music.

That said, tapping in to music isn't as complicated as tech companies make it sound. We derive the most benefits by listening to songs we love and choosing the right tempo for the task at hand. Do we really need an app for that? No, said Robert Zatorre, the neuroscientist who contributed to the discovery that music triggers dopamine. As for AI that predicts our pleasure response, "Spotify already has that nailed down pretty well."

But in the near future, he added, technology-driven music might offer new treatments for people coping with severe depression, substance abuse, or Parkinson's disease. People with these disorders tend to have poor regulation of the dopamine system in the brain. To train this system, Zatorre has teamed up with Israeli researchers in a project called Music to My Brain, using computer-generated (or "generative") music and a technique called neurofeedback. While listening to music, patients are hooked up to EEG machines that track activity in the brain's reward regions. Their brainwaves signal to a computer to make the music louder or softer, faster or slower, depending on their pleasure levels. The better the music sounds to the patient, the more it activates their reward system, Zatorre explains, which in turn makes the music sound even better, and so on. "It's a positive feedback loop."

The goal is to regulate the dopamine system, thereby reducing symptoms of depression or Parkinson's, or cravings for drugs and alcohol. Patients might need regular music sessions to maintain any gains, Zatorre said, but "not every waking moment of every day." If this approach works as well as he predicts, generative music could be a game changer.

In the meantime, most functional music apps amount to digital frills. Few appeal to me, because plain old music does the trick on its own.

To tell the truth, though, I used to give music listening far too little credit, whether recorded or live. It was not until my late thirties, during one of the toughest years of my life, that I fully grasped just how potent it could be.

MY BABY WAS BORN with a dimple in one cheek and a set of lungs that would soon produce blood-curdling screams. He screamed in his stroller. He screamed on his play mat. He screamed in his car seat, baby swing, crib. I tried every baby-soothing technique in the book, but at night he would only settle for an hour or two at a stretch. "Cry it out" didn't work: a six-night trial ended in even less sleep for baby and me.

During the day, I couldn't "sleep when the baby sleeps" because mine would only nod off if I trudged through Vancouver's rainy streets with him strapped to my back. I became the walking dead—bone-achingly, brain-numbingly sleep-deprived. My husband was working seven days a week to keep his fledgling business afloat, and neither of us had family nearby who could help. Babysitters couldn't stand the crying.

That left me.

I didn't sleep for five hours in a row until our son was six months, and this blessed record lasted all of one night. In my altered state, I worried that I'd never get back to my journalism job. I could barely get through each day.

Nine months in, on yet another bleary trudge through the city, a flyer on a lamppost caught my eye: African Movement Workshop. The woman in the photo glowing with her drum looked more vibrant than I'd felt all year. My first break from mothering was long overdue, but I didn't ask myself if I'd be better off with a therapy session or a boozy lunch with friends. I whipped out my cell phone and signed up on the spot.

In the carpeted office space, musicians with goblet-shaped djembe drums got started in a leisurely groove. The workshop leader asked the rest of us to stamp our feet to the bass tones and clap our hands to the drums' staccato beats. Then she showed us how to move across the room in small groups to experience "the collective unity" of an African village. Nothing in this boxy room summoned a rural village for me. Shuffling across the floor with half a dozen strangers, I felt silly, awkward. *What am I doing here?*

Next, the leader told everyone to find a spot on the floor and "move to the music in your own way... whatever feels good." As the drummers picked up the pace, a woman across the room pressed her palms together and wiggled in a kind of snake charmer's dance. Someone else lay on the floor vigorously shaking their arms and legs, while the guy beside me whirled like a windmill. Overwhelmed by the pounding drums and chaotic moves, I inched away, eyes closed, head reeling.

Moments later, a loud rattling jolted me. Peering through my eyelashes, I saw the workshop leader beside me holding a dried gourd the size of a beach ball strung with red and turquoise beads. The gourd resounded and crackled as she thumped its base and slapped the beads around the calabash. She circled me as I stood there in a daze, feeling the beats penetrate my skull and ricochet in my head, splintering my thoughts. For a moment, nothingness. Then, from the far end of the room, the warm tones of a marimba rippled through the air. The gentle currents bobbed around me, mellow and cheery, and I had an almost physical sensation of being lifted by the exuberant melody. After months of shouldering my baby, it was as if something was carrying me—weightless, floating, jubilant. I burst into tears of relief.

I'd been uplifted by music before, but never like this. An uncanny feeling had permeated my entire body, triggering a cathartic release that would stay with me for weeks. That day marked the turning point in my postpartum struggles. Despite my ongoing sleep debt, I felt lighter, more able to cope. How could this be?

As a journalist, I was trained to question all things woo-woo, but I didn't need a before-and-after brain scan to know something had shifted in me. A neuroscientist might explain it as surges of dopamine and endogenous opioids, the brain chemicals that bring pleasure and numb pain. But what was it about the rattling of the gourd and the timbre of the marimba? Could any music have done the trick, or any break from mothering?

I wondered if talk therapy would have given me the same wave of relief. My geeky side wished I could have done a controlled experiment. The rest of me vowed to stop underestimating what music could do.

— 7 —
Bad Vibrations

WHEN MY SON was a toddler, I didn't have much time for what the wellness gurus called "self-care." Deadlines at the *Globe* had ramped up. On top of writing in-depth health features, at 3:30 p.m. every workday, I had to pitch a "hot button" blog story and have it ready to go live at 5 p.m. I lived in fear of blowing the deadline or letting an egregious error slip through. Meanwhile, at home, I picked up the slack as Scott's sustainability-minded engineering business went gangbusters. Unlike my mom, who left me and my siblings mostly to our own devices, I spent every spare moment with my child, reading out loud, hosting playdates, or sewing curtains for a homemade puppet theater. I felt lucky to have quality time with him. Still, I was tired.

One afternoon I carved out an hour for a rare massage in a studio with low lighting and a pine scent. Even though my face was scrunched into a doughnut-shaped head support, I could imagine basking in a forest on a soft bed of moss. Warm hands pressed into my back, coaxing my muscles to release. Aaahh.

The soundtrack switched from bird calls to undulating arpeggios in G. Instantly, I recognized the opening of Bach's *Cello Suite no. 1*: fluid, tawny, beautiful—and the last thing I needed. My throat constricted. It was like hearing the voice of an old flame who had broken my heart. I stirred on the table, unable to bear it. "Can you please turn it off?"

The massage therapist asked if I was okay, and I felt obliged to explain. She patted my shoulder, comforting me. But as the session continued in silence, my mind went down the well-worn rabbit hole of soft-tissue injuries and futile striving. I left the massage studio feeling more agitated than when I walked in.

To this day, classical music fills me with something like dread. Nothing has changed about the music I once loved; nevertheless, mixed emotions and layers of meaning have turned symphonies and concertos into auditory kryptonite. Bad vibrations, you might say.

I have a theory about my aversion, which I'll get into later. My point here is that music is seldom neutral, even at volumes and frequencies that don't harm our eardrums. Tunes trigger knee-jerk reactions in us, good and bad. Lab studies confirm this: when we hear music we dislike, brain connectivity in our default network—a region involved in empathy and self-awareness—drops.

Most of us can name certain songs that bring back bad memories, but the ones we like can drive us bonkers, too. Up to 98 percent of people have been bothered by an earworm, also known as "sticky songs," a type of involuntary auditory cognition. Earwormy tunes tend to have a fast tempo and catchy melody. In a 2016 study from the American Psychological Association, top sticky songs included Lady Gaga's "Poker Face" and Journey's "Don't Stop Believin'." (The antidote: chewing gum. In experiments at the University of Reading, England, jaw movements helped short-circuit involuntary auditory loops.)

Beyond earworms, though, can music cause true suffering? This question doesn't get much attention because we tend to romanticize music, convinced of its intrinsic goodness. In the memoir that inspired *The Sound of Music*, Maria von Trapp, the Austrian nanny turned naval commander's wife, called music "a magic key, to which the most tightly closed heart opened." Days after the assassination of U.S. President John Kennedy, the conductor Leonard Bernstein declared that a musician's best response to violence is "to make music more intensely, more beautifully, more devotedly than ever before."

Yet music has always had a Janus face. As the writer Alex Ross chronicled in *The New Yorker*, the siren songs of Greek mythology lured sailors to their death. The Israelites blasted trumpets as psychological warfare in the biblical Battle of Jericho. During a more recent siege, in 1993, the U.S. Federal Bureau of Investigation used music as an intimidation tactic for nearly seven weeks, pummeling a cult compound in Waco, Texas, with nonstop bugle calls, Tibetan chants, vintage Christmas carols, and Nancy Sinatra's "These Boots Are Made for Walkin'." Then there's the chilling scene from *Apocalypse Now* when American choppers gun down Vietnamese villagers to the theme of Wagner's "The Ride of the Valkyries." Blared by the military, opera becomes menacing.

I'll admit that after years of ruminating over my relationship to the cello, the dark side of music fascinates me. My curiosity about the potential harms of music might have to do with its double-edged role in my life: joy and obsession, pleasure and pain. In my work as a health reporter, I've learned that anything potent enough to offer benefits, from blood thinners to chemotherapy, generally carries risk. So, if music has drug-like effects, can there be too much of a good thing? Is death metal the scourge it's made out to be? If music heals, can it harm?

Ross pondered similar questions in his *New Yorker* article "When Music Is Violence." If music can change our perceptions and emotions, he reasoned, then "it must also be able to act destructively."

IN A BUSTLING Austrian town in 1901, a twelve-year-old boy took a seat in a theater near the banks of the Danube. Thin and lank-haired, he watched in awe as a legendary knight appeared in a boat drawn by swans. Wagner's *Lohengrin*—source of the "Here Comes the Bride" song—was the first opera the boy had ever seen. "I was fascinated at once," he wrote years later in *Mein Kampf*. "My youthful enthusiasm for the Bayreuth Master knew no bounds."

Adolf Hitler claimed that as a teenager in Vienna, he went without food to see operas. In Wagner, he found a powerful muse for his vainglorious fantasies and nascent racial hatred. Although the composer died before Hitler's birth, Wagner had fanned the flames of anti-Semitism in Europe. In a hate-laden essay aimed at Jewish people, Wagner insisted they must renounce their distinct culture and religion to assimilate with his kind. Citing the myth of the Wandering Jew, doomed never to belong, Wagner wrote: "Only one thing can redeem you from the burden of your curse... going under!"

Hitler never mentioned Wagner's anti-Semitism publicly, but given his lifelong obsession with the composer, he was no doubt aware of his views. In Hitler's eyes, Wagner's operas personified Teutonic destiny and might. "The reason we perceive the artist Richard Wagner as being great," Hitler declared at a Nazi meeting in 1923, "is because he represented heroic folklore, Germanness, in all his works."

Hitler knew that facts and ideology wouldn't be enough to sway the masses, so he harnessed the unifying power of music to capture hearts and minds. In 1933, the Nazis celebrated the first months of the Third Reich with a five-hour-long performance in Berlin of Wagner's *Die Meistersinger von Nürnberg* (The Mastersingers of Nuremberg), which they regarded as "the most German" of all operas. In one scene, a town clerk named Beckmesser—almost certainly a Jewish caricature—is beaten up by a cobbler's apprentice in a jealous rage.

Hitler's propaganda minister, Joseph Goebbels, banned "degenerate" music such as jazz, along with works by Jewish composers including Mendelssohn. A year before Hitler made it illegal to listen to foreign broadcasts, Goebbels declared radio "the most influential and important intermediary between a spiritual movement [Nazism] and the nation." To increase the spread of Nazi-approved music and ideas, the regime mass-produced cheap radios known as the "people's receiver" (later models had a tiny eagle with a swastika on either side of the tuner display). Starting in 1934, radio ownership soared from a third of German households to nearly two-thirds by 1938. Day in and day out, Nazi radio broadcasts played only "good German music," from the waltzes of Richard Strauss to German folk tunes that conjured nostalgia in the collective memory.

Hitler's strategy to weaponize music extended to the death camps. At Sachsenhausen, north of Berlin, guards humiliated prisoners by ordering them to sing and dance to Nazi songs. "Those who didn't know the song were beaten," said Eberhard Schmidt, a former prisoner. "Those who sang too softly were beaten. Those who sang too loudly were beaten."

At Auschwitz, Primo Levi witnessed prisoners marching to "Rosamunde," also known as the "Beer Barrel Polka." Sung in America by the Andrews Sisters, this cheery tune was twisted into a grotesque mockery of suffering. Meanwhile, inmate musicians were forced to perform during executions, in front of the entire camp, or play near the gas chambers at the whim of SS officers. "The beating of the big drums and the cymbals reach us continuously and monotonously," Levi wrote. "We all feel that this music is infernal."

I've performed some of Hitler's beloved Wagner, including "The Ride of the Valkyries," named for the mythical Norse maidens who flew through the air carrying slain warriors to Valhalla. The triumphant triads of the "Valkyries" are thrilling to play. But when I think of this pleasure, and the hours I spent analyzing Wagner's harmonic progressions for university assignments, I am repulsed. Not one of my professors or music-history textbooks ever mentioned the Nazi angle. On Hitler's fiftieth birthday, Wagner's descendants gifted him with a bundle of the composer's original scores. At the end of the war, Hitler clutched his precious Wagner operas—including *The Valkyrie*—in the Berlin bunker where he shot himself.

Wagner fans may argue there's nothing inherently evil in the notes themselves, but I can no longer separate musical genius from sinister history. Wagner's music, for me, will be forever tainted.

MUSIC CAN FRIGHTEN US, bore us, annoy us, distract us, or fill us with disgust. "That's not music," we complain, "that's noise!"

But what's the difference—and who decides?

Some noises will set anyone's teeth on edge, like the buzzing of a mosquito or the whining of a child. Once we call it "music," though, consensus falls apart. The line between music and noise is so blurry that the composer John Cage refused to draw one. In his most provocative work, entitled "4′33″," musicians do nothing but sit onstage for the time span in the title. From wailing sirens to coughing in the audience, ambient sounds, to Cage's ears, all counted as music.

The British anthropologist John Blacking made a sharper distinction, describing music as "humanly organized sound."

Was Cage's "4′33″" organized enough to qualify? That depends on who's listening. While scholars haggle over definitions, most of us think of music the way lay people describe art: "I know it when I hear it." But whether we experience music as "good vibes" or an affront to our ears is often a matter of taste. As the musicologist Felipe Trotta points out in his book *Annoying Music in Everyday Life*, "bad music" generally applies to sounds beyond our control. "Being exposed to music we do not choose means being forced to vibrate according to it," he writes, "even if the set of ideas that are shared hurts our moral sense."

As we've seen, music can alter our heart rate, immune markers, and even our brainwaves. If we don't like it, it's an auditory assault. In Modesto, California, a 7-Eleven franchise found an effective way to keep teenagers from loitering: blasting classical music. Using the same strategy, New York's Port Authority played Handel through loudspeakers, while transit hubs in Portland, Oregon, piped tunes from operas such as *Carmen*. One person's auditory ambrosia is another's infernal racket.

Musical preferences, writes Trotta, come from a "lifelong construction" of memories, ideas, feelings, and experiences of belonging. But if the noxious effects of music merely depend on taste, how harmful can they be? This enigma has puzzled great minds since the days of the lyre.

PLATO WAS DETERMINED to weed out bad vibrations. In ancient Greece, bursts of experimentation had given rise to musical scales, or modes, named for neighboring peoples: Ionian, Dorian, Phrygian, Lydian, Aeolian. To Plato, they must have sounded like the "Thong Song." In his *Republic*, he banned the Lydian mode for encouraging drunkenness and found fault with most of the others. Only the Dorian, dignified and "without frills," and the Phrygian, a useful stimulant for battle, belonged in his utopian city-state.

Now we fast-forward to the Enlightenment, when hand-wringing over the "wrong" songs gained scientific gloss. Taking cues from René Descartes's mind-body divide, seventeenth-century scientists described body parts in mechanical terms. Nerves became fluids, writes James Kennaway, a historian of medicine at Durham University, England. Music was a direct-acting "nerve tonic." In the right doses, it was healthful. But in patients with illnesses such as "inflammatory fever," cautioned a physician who hobnobbed with the Mozart family, music could have "damaging consequences."

Growing fears about music coincided with booming piano sales and a mushrooming middle class. In 1824, a wellness book called *The Family Oracle of Health* underlined "the bad effects of music on the nerves" and claimed a single performance of a Rossini opera had caused "more than forty cases of brain fever, or violent convulsions." Music needed a warning label.

While the "threat" applied especially to the gentler sex, women of the day enjoyed living on the edge. Listening to the Hungarian virtuoso Franz Liszt tickling the ivories was said to cause fits of delirium. But just like Beatlemania a century later, raving at a Liszt concert had social cachet. "So fervid was feminine admiration," wrote the music historian Percy Scholes, that if Liszt dropped his handkerchief, "it was torn to pieces as 'souvenirs.'"

Girls gone wild, Victorian-style.

Long before the notion of "subliminal messages" in heavy-metal albums, nineteenth-century composers toyed with a kind of musical mind control. Wagner used recurring themes called leitmotifs to jog the opera listener's memory of everything from a giant's castle to a dwarf's curse on a ring. He often embedded these musical signals in ways the casual listener would never detect. Before the 1865 debut of *Tristan and Isolde*, Wagner even speculated his latest opera might drive people insane. Mark Twain attended *Tristan* at the Bayreuth Festival in 1891, and reported feeling "like the sane person in a community of the mad." (But in the same breath, he described it as "one of the most extraordinary experiences of my life.")

IN THE TWENTIETH CENTURY, a new genre of music reared its head, one so debauched it was said to atrophy the brain. Under its "demoralizing influence," wrote the music educator Anne Shaw Faulkner, listeners "are actually incapable of distinguishing between good and evil." Faulkner wasn't talking about death metal or profanity-laden rap. Writing in the *Ladies' Home Journal*, she was describing jazz.

Her alarmist screed, published in 1921, was echoed sixty years later amid fears about heavy metal. A bill put before the California Assembly in 1982 claimed that albums by bands from The Beatles to Styx contained secret messages when played backwards. These "back-masked" messages, argued California legislator Phil Wyman, could "turn us into disciples of the Antichrist." Moral panic grew so intense that former Second Lady Tipper Gore launched a campaign to have "parental advisory" stickers added to music albums. Rock stars including Dee Snider, lead singer of Twisted Sister, testified in U.S. Senate hearings to defend their craft.

Gore's pearl-clutching over Twisted Sister now seems quaint. Today's death metal bands have names like Cannibal Corpse, and scream lines about crushing people's heads until the brains drip through. Millions of listeners worldwide are drawn to songs that shriek about murder, torture, infanticide, and rape.

To better understand the attraction, Australian psychologists asked forty-eight death metal fans and ninety-seven non-fans to listen to songs in this vein. Not surprisingly, music with violent themes left non-fans feeling "tense, afraid, and angry." For death metal fans, though, the same songs had the opposite effect: devotees reported increased feelings of personal power, inner peace, and joy. One listener described death metal's appeal as having "something to do with the primal scream in us. It's a release, accepting and empowering."

The psychologists concluded that death metal fans, like horror buffs, are able to separate fantasy from reality. The shocking lyrics repel outsiders, reinforcing a sense of shared identity. For fans, death metal is "feel-good" music.

Does this mean horrifically violent lyrics set to music are harmless? I'm not convinced. Music has strong social-bonding properties (see chapter 8) and it's easy to imagine how a shared attraction to hate songs could stoke violence. Tzvi Fleischer, an Australian-based writer and editor, has described racist music, spread through the Internet, as "perhaps the most important tool of the international neo-Nazi movement to gain revenue and new recruits."

To me, that's sinister enough. Military forces, however, have taken music to barbaric lengths.

A PRISONER SHIVERS alone in a dark cell, hands and feet shackled to the floor. Headphones strapped to his ears block out all sounds other than a cloying voice singing the *Barney* theme song hour after hour, until it feels like it's drilling into his skull. The inmate has no idea how long this barrage has gone on, because he can no longer think straight. Music, played ad nauseam, is making him lose his mind.

Torture by Barney tune sounds more like a punchline than a punishment: "Beware the purple dinosaur. Resistance is futile." In the War on Terror, though, Barney music was part of "enhanced interrogations" by the American military. Teams of U.S. Psychological Operations (PSYOPS) soldiers bombarded detainees with heavy-metal music and repetitive children's songs, including Barney. "Trust me, it works," a U.S. operative in Iraq boasted to *Newsweek*. The idea, said Sgt. Mark Hadsell, was to break a prisoner's spirit using culturally offensive music. "These people haven't heard heavy metal before. They can't take it," he said. "If you play it for 24 hours, your brain and body functions start to slide, your train of thought slows down and your will is broken. That's when we come in and talk to them."

Ruhal Ahmed, a British citizen, was held without trial for more than two years at a U.S. detention center in Guantánamo Bay, Cuba. He described the musical assault as worse than physical torture: "Once you accept that you're going to go into the interrogation room and be beaten up, it's fine," he said, after his release without charge. "You can prepare yourself mentally." But music at maximum volume "makes you feel like you're going mad," he explained. "After a while, you don't hear the lyrics at all—all you hear is heavy, heavy banging."

Donald Vance, a U.S. Navy veteran from Chicago, detailed similar abuses during his wrongful imprisonment in Iraq. Vance was working for a security firm in Baghdad when he was captured by U.S. officials, who suspected him of colluding with Iraqi arms dealers. At the age of twenty-nine, he found himself on the wrong side of a steel door in a U.S. maximum security center in Baghdad: Camp Cropper.

Guards left fluorescent lights on in his cell 24-7, and had "goddamn blaring music" playing nonstop. Songs in heavy rotation included Queen's "We Are the Champions." As time wore on, hearing songs he liked—songs he would play at home—"began destroying me." After his release, Vance said his ninety-seven days at Camp Cropper left him with chronic depression, nightmares, and paranoia.

Amnesty International has denounced prolonged and involuntary exposure to loud music as a form of torture. But how to pry apart the effects of music from the impact of loud noise, sleep deprivation, and solitary confinement?

If music can destroy minds, its use as psychological warfare would be unlawful under the 1949 Geneva Conventions, which prohibit torture and "outrages upon personal dignity." During the War on Terror, though, the Bush Administration decided such treaties did not apply to suspected al-Qaeda and Taliban members detained outside the United States.

At Guantánamo Bay, U.S. interrogators tormented prisoners with any music they had on hand: gangsta rap, Britney Spears, Christina Aguilera. One inmate, Detainee 063, was subjected to this cruel and unusual punishment nearly forty times in forty-nine days. The prisoner was twenty-three-year-old Mohammed al-Qahtani, the alleged "twentieth hijacker" involved in the September 11 attacks. Guards made him listen to sexually charged pop songs, violating Islamic fundamentalist beliefs forbidding Western music. In one entry from the official log of interrogations, "Detainee broke down crying and asking [Allah] for forgiveness." His condition declined so rapidly that he was put under a doctor's care, but even then, music was played to "prevent detainee from sleeping." In a report to the Pentagon about inhumane practices at Guantánamo Bay, FBI agents described al-Qahtani's behavior as "consistent with extreme psychological trauma."

Whether music contributed to his break from reality remains unclear (al-Qahtani did have a history of mental breakdowns before his detention). Speaking to the BBC in 2003, Rick Hoffman, vice president of the U.S. PSYOPS Veterans Association, insisted the use of music in interrogations would have "no long-lasting effect." But Michael Peel, a doctor working with the Medical Foundation for the Care of Victims of Torture, disagreed. "Music is used to make the detainee aware that he has no control over what's going on in any of his senses," Peel said in a 2008 BBC interview. Such treatment "eventually dehumanizes people."

With music, as with sex, consent matters—and context is everything. Gene Kelly's "Singin' in the Rain" has not a hint of savagery, yet I can't hear this song without flashing back to the scene in *A Clockwork Orange* where a vicious teenager rapes a woman after beating her husband to this gleeful tune. As one movie buff described it, the mix of sexual violence with a family-friendly song "makes your brain short a fuse," entangling wires that shouldn't get crossed.

Neuroscientists have demonstrated that our brains entrain to musical rhythms. What happens to someone whose brainwaves are stuck in the same oscillations for hours on end? To my knowledge, such an experiment using brain scanning hasn't been done, for ethical reasons alone. Yet there's no doubt that songs penetrate deep into the emotion centers of the brain, in ways that bypass our conscious volition and choice. We know that music can alter core physiological systems in the body—and a prisoner in a cell block can't make it stop.

Beyond prison walls, though, some people will fight tooth and nail to keep the music playing.

ROGER TULLGREN TOOK A SHINE to heavy metal music at the age of six, when his older brother came home with a Black Sabbath album. Years later, at forty-two, the Swedish metalhead claimed he couldn't stop himself from missing work to see touring acts and insisted that his "handicap" interfered with his ability to hold a job. Remarkably, after psychological evaluation, the employment center in his hometown of Hässleholm granted him disability benefits.

Tullgren got part-time work as a dishwasher, on condition he could listen to bands like Iron Maiden on the job. The employment center subsidized his salary for five years, but then cut him off. "I'm still very addicted," Tullgren lamented in a 2013 follow-up article. "Metal controls my life."

Even in semi-socialist Sweden, his case is exceptionally rare. Disability benefits generally require a medical diagnosis, and in the diagnostic manual of psychiatric disorders, "music addiction" is nowhere to be found. Should it be added to the next edition?

I've never seen any research on treatment for a hip-hop habit, but music does trigger the release of dopamine, the brain chemical underlying all addictions. What's more, music may amplify the effects of drugs and alcohol. In an Australian study of inpatients receiving treatment for substance abuse, nearly half identified a specific song or type of music that cued cravings for their drug of choice—a kind of Pavlovian response. Patients described a synergistic effect: music intensified their emotions while under the influence, and on the flip side, drugs and alcohol deepened the emotions they felt from music. Sex, drugs, and rock 'n' roll.

Substance abuse aside, though, is "music addiction" for real?

The neuroscientist Robert Zatorre, a world expert on music and dopamine, told me he's never heard of any harmful effects from music, "even for someone who is sort of obsessed with it, so I'm skeptical of the idea that it has real addictive properties."

But Kevin Kirkland, a music therapy professor at Capilano University in North Vancouver, said he has observed problematic music behaviors in his work at the Burnaby Centre for Mental Health and Addiction. Although music itself didn't cause psychiatric problems, he added, some clients with schizophrenia, depression, bipolar disorder, or PTSD use music as a way of "sublimating" unwanted emotions or impulses. Instead of expressing themselves through language and social relationships, they self-medicate by playing music for hours and hours a day.

If music dependence exists, it would qualify as a "behavioral addiction" similar to problem gambling. To qualify as an addiction, however, music-related behaviors would have to have damaging effects on a person's health, livelihood, or social relationships for more than a few days or even weeks. The difference between musomania (music obsession) and addiction, said Mark Griffiths, a psychology professor and director of the International Gaming Research Unit at Nottingham Trent University, "is that healthy excessive enthusiasms add to life, whereas addiction takes away from it."

Very few studies have looked at "problem" music habits. In one exception, researchers at Northeastern University measured "maladaptive music listening" by altering questions from an alcohol screening test designed to identify addictive behaviors and negative impacts in daily life. In this study, nine out of fifty amateur musicians met the criteria for music dependence, compared to a single individual from the same number of non-musicians.

None of this worried me, because I've never gone overboard with music listening. But then I came across a 2017 report noting that "some musicians either continue to practice through practice-induced pain or have psychosomatic disorders at deprivation [of practicing]." The study involved twenty-five classical musicians from German conservatories. Using a scale adapted from exercise addiction, the researchers identified three of these students as being "at risk for dependence" and twenty as having possible symptoms of "practice addiction" (which the study acknowledged as a hypothetical condition).

In a box in my garage, I have notebooks filled with obsessive lists of when, what, and how much I practiced, down to the minute. I remember beating myself up when I missed a session, like an exercise addict who misses a workout. Was I hooked on the lush sounds of the cello, or simply caught in a hamster-wheel approach to practicing? I never asked myself this question at the time. Denial came easily because in university, obsessive practicing was the norm. In the hallway outside the practice rooms, my peers and I would brag about how many hours we'd put in.

Later, in my post-cello years, my mother described me as going through a kind of withdrawal. "All those vibrations surrounding you every day, suddenly gone," she said. "How could you not be affected by that?"

QUITTING THE CELLO had set me adrift professionally, socially, and emotionally. As I struggled to create a new identity, I didn't just deprive myself of the rich vibrations my mother rhapsodized about. I also developed an aversion to anything composed before the 1940s.

As a teenager, I listened to *Gilmour's Albums* on CBC Radio and got a kick out of guessing the classical composer or piece, playing my own version of *Name That Tune*. Now the sounds of classical instruments, especially the cello, make me feel uneasy and tense. They're tough to avoid, though: in movie soundtracks and TV series, cellos accompany love scenes, tragedies, sexual assaults. Pleasure and pain.

My drastic change in musical preference is rare, as most people stick with the music styles they discovered in their teen years. In a 2020 survey from Nielsen Music, 87 percent of American music consumers said they seldom strayed from the music they normally listened to. "Our musical taste is greatly influenced by what we hear between the ages of twelve and twenty," said the neuroscientist Daniel Levitin. "Part of it has to do with brain development and the pubertal growth hormones and the way the brain is maturing."

People with a personality trait psychologists describe as "openness to experience" may be more likely to expand their musical tastes. "It's something you can make an effort to do, go out and hear new concerts," said Levitin. "Whether it will take or not depends on your brain and your genetics."

In my case, though, a visceral reaction to stringed instruments might have to do with what I call "musical PTSD."

Post-traumatic stress disorder is a serious medical condition, and I don't mean to trivialize it by comparing my issues with those of war veterans, cancer patients, or survivors of sexual assault. I might not even use the term if I hadn't spoken with the neurologist Barry Bittman (see chapter 4). As a board-certified pain management specialist, Bittman has treated a number of professional musicians who could no longer touch their piano, cello, or timpani. One pianist had such debilitating pain that he couldn't drive his car. While their distress manifested as physical symptoms, he said, these musicians also suffered from a host of psychological issues: Perfectionism. Anxiety. Aversion to an instrument they had loved. Bittman realized this was "very similar to a post-traumatic stress disorder."

Unlike true PTSD, none of my experiences completely debilitated me. Even so, I can see parallels with the Mayo Clinic's description of the disorder. People with PTSD tend to startle easily; have outbursts of anger, shame, and guilt; and avoid things that remind them of traumatic events. Intrusive memories or flashbacks torment them in their waking hours or dreams. In between triggers, a sense of numbness sets in, combined with feelings of hopelessness and low self-worth.

Although my "symptoms" were milder by far, they still persisted for years. I had nightmares about being in a room with my first cello teacher or choking onstage. In my late twenties, despite needing the cash, I dropped a gig as a classical-music critic because I was triggered by the sounds and didn't like the nit-picking they brought out in me. Meanwhile, I shied away from string players I knew from my cello days, convinced they judged me for quitting. Sometimes I still feel residual shame and guilt.

I didn't realize how many ex-musicians had issues until I read a *New York Times* article entitled "The Juilliard Effect: Ten Years Later." Graduates told the writer Daniel J. Wakin about their shattered dreams and identity problems. I nodded at every line. Fewer than a quarter had steady orchestra jobs. Others eked out a living from sporadic gigs or had left the field entirely. Some had problems with drugs and alcohol. One had committed suicide.

Their trajectories paralleled those of my peers. A violist I knew in university got involved in a cult and took his own life. A cellist a few years older than me died of drug overdose. Another exceptionally talented string player took swigs from a flask before we went onstage. The writer Hugh Morris exposed this all-too-common dependency in his 2021 article, "Confronting Classical Music's Alcohol Problem." Musicians surveyed in the UK consumed at least a third more alcohol on average than the typical drinker. But in the "prim" and "frankly

Victorian" classical-music world, Morris wrote, "any attempt to assess what is a serious, deep-rooted problem is seen as sucking the little joy out of what is a universally difficult profession."

None of these "side effects" come from the music itself, let alone specific rhythms or frequencies. Most musical harms, I've concluded, manifest when a human pleasure is mixed with coercion, intense pressure, or acts of cruelty. There are exceptions, though.

In pseudo-medical interventions using music, sometimes the loftiest of intentions end up doing more harm than good.

WHEN A MANHATTAN ONCOLOGIST advised cancer patients to listen to singing bowls, no one could blame them for giving it a try. After all, Mitchell Gaynor was a former clinical assistant professor at the Weill Cornell Medical College, and the former director of medical oncology at its Center for Integrative Medicine. Why would any patient doubt his advice?

In his book *The Healing Power of Sound: Recovery From Life-Threatening Illness Using Sound, Voice, and Music*, Gaynor described an encounter in New York City with a Tibetan monk. The holy man played a singing bowl that touched Gaynor to his core "in such a way that I felt in harmony with the universe." Soothing music, Gaynor wrote, can realign the body with the essence of all creation.

It's such a compelling idea. From jitterbug mitochondria to planetary bodies pirouetting around the sun, everything in the universe is made up of vibrations. One theoretical physicist, Brian Greene, proposed that all matter, down to the smallest subatomic particles, comes from tiny strings of energy that vibrate in "a kind of cosmic symphony."

Gaynor, though, turned a metaphor for the infinite into a medical prescription. "I believe that sound can play a role in virtually any medical disorder," he wrote, "since it redresses imbalances on every level of physiologic functioning."

But the truth is, we cannot vibrate ourselves free of cancer. Life-threatening illnesses don't work that way. Much as I respect music's therapeutic potential, I also believe that too much faith in "healing vibrations" can fill desperate patients with false hope. A Calgary businessman who worked with a relative of mine made the trip to New York City for sound healing at Gaynor's center. Musical vibrations couldn't cure his terminal cancer.

Nor could the physician heal himself.

Gaynor relied on chanting and singing bowls as part of his personal wellness plan, but in 2015, his body was found in his country home in upstate New York. He was fifty-nine, and died of suicide. Did his faith in sound healing prevent him from reaching out for psychological help that might have saved his life? No one can say for sure.

But as it turns out, "Tibetan singing bowls" have nothing to do with traditional Tibetan healing. Their use as a sound-healing tool originated in the West. Commenting on singing bowls in the *Toronto Star*, Tenzin Dheden, a member of the Canada Tibet Committee, implored people to "kindly stop mythologizing and exoticizing Tibetans, and leave us out of your pseudo-scientific New Age nonsense."

"Sound bathing" is another healing strategy that's been filled with hot air. Whereas a music therapist has years of university training before certification in their field, anyone can hang out a set of wind chimes and call themself a "sound healer." The trend has people lying on yoga mats as self-styled "healers" draw vibrations from metal gongs, chimes, and quartz-crystal singing bowls.

America's largest sound-bathing series packed more than 1,200 souls into San Francisco's Gothic-Revival-style Grace Cathedral. The event's founder described it as an antidote to stress: "People are looking for some sanctuary, some place to go to unplug from the daily madness." Nothing wrong with that. Any relaxing activity can lower cortisol levels, thus supporting mental and physical health. But instead of framing the benefits this way, too many "sound healers" spout hokum about "massage on the cellular level" and claim this "ancient modality" holds mystical healing powers.

When new-age healers persuade cancer patients to forgo life-saving treatments, they can do real harm. Within five and a half years of a cancer diagnosis, Yale University researchers reported, the risk of dying increased by 150 percent for patients who opted for alternative remedies alone compared to those who chose chemotherapy, radiotherapy, or surgery.

While the study didn't single out singing bowls, sound healers do target cancer patients. Promos for singing-bowl workshops often parrot a quote attributed to the French composer and self-described "bioenergetician" Fabien Maman: "Cancer cells cannot maintain their structure when specific soundwave frequencies attack the cytoplasmic and nuclear membranes." This claim even holds an iota of truth. Beams of ultrasound—at ten thousand times the power used in pregnancy—can heat cancer cells to above 55 degrees Celsius, zapping them dead. The catch: vibrations from tuning forks or singing bowls could never approach ultrasound, and if they could, they'd be dangerous in unskilled hands.

This is a standard new-age tactic, of course: latch on to something factual, add scientific jargon, and then run with it.

To be fair, I can understand the appeal of sound bathing. Unlike the glut of digital meditation apps, sound bathing is live, improvised, and draws people together. As someone who had a postpartum epiphany in an African music workshop, I can hardly scorn people who swear it takes a load off.

I gave sound bathing a try (for research purposes) while vacationing on Hornby Island, a bastion of hippiedom off the coast of British Columbia. At the hobbit-style community hall, entered through a hollow tree stump, a bulletin board pointed me to a sun-washed cedar grove. Inside the yoga studio, gleaming tuning forks and "crystal alchemy bowls" were arranged on the floor like a sacred mandala.

Patchouli and sandalwood perfumed the air as the sound healer, a soft-spoken woman in her thirties, gave a preamble about our "natural attraction" to harmonic patterns. Any departure from these harmonies, she continued, "creates dissonance, chaos." *But,* my mind argued, *dissonant sounds exist in nature, too. Yin needs yang, right?*

I lay on the floor surrounded by strangers, all heads pointed towards the musical instruments. A metallic chime jingled, followed by the murmurings of a tuning fork and the deep resounding of a gong. There was something comforting about communing with music in a group, without any pressure to talk or interact. Little by little, I stopped naming each instrument in the shimmering wash of sounds. Niggles gave way to a lulling daydream.

"Now," said the sound healer, drawing the session to a close, "bring your attention back to your breathing." Even if I didn't buy her esoteric claims, I had to admit to being chilled out. I felt woozy, almost drugged. No one had kneaded my muscles or touched my skin, yet my body was far more relaxed than it had been after cello music took over my massage. This time, no bad memories, no baggage. None of the sounds reminded me of anything else.

As I left the yoga studio, I reminded myself that I didn't need to pick up the cello again to make music. Anyone could do it, including me, even if it's as simple as the reverberating ring of a gong.

— 8 —

All Together Now

EACH TIME I dropped my palm near the center of the drum, the warm goatskin bounced back, as if alive. *Goon!* The bass note exploded from the hollowed-out tree trunk between my knees, shooting into the floor and the air around me. *Goon!*

Such a dramatic sound, for so little effort. *Goon! Goon! Goon!* I could do this all day, but my initiation into hand drumming started as a Tuesday-night thing. After returning to work from maternity leave, I enrolled in drum lessons with Navaro Franco, leader of the workshop that had helped me through my postpartum angst. A charismatic woman in her late forties, she had a veil of dark hair and a dancer's physique. She had performed for years with a disciple of Babatunde Olatunji, the Nigerian master widely credited for turning North Americans on to hand drumming. Although she wasn't African, she knew the pulses in her bones.

Each lesson had a ritual feel. At the center of the circle, candles flickered like a primordial campfire, casting light on a frond of cedar, a glossy feather, a river-tumbled stone. As a warm-up, we traced circles on the drums with our hands. "Feel the skin of your fingertips touching the drum," the teacher said. "Relax your shoulders. Let each breath fill your belly." Our bodies began to synchronize before we played a single note.

The teacher introduced the rhythms by singing the different "voices" of the drum: "Goon Goon" (or "Gun," for bass), "Go Do" (tones), "Pa Ta Pa Ta!" (slaps). She had us vocalize the beats of each new pattern, clap them with our hands, or step them with our feet. We absorbed them in our bodies, not just our brains. No more Cartesian mind-body divide. She encouraged us to sense the pulses in relation to our breathing, heartbeat, vocal cords, gait, and then extend this awareness to broader cycles—from the undulations of ocean waves to the revolutions of the seasons, the Earth, and the moon. She described drumming as a "tribal technology," an ancient strategy for balancing body and brain. My cynical side scoffed, but the rest of me wanted more. Drumming with her made me feel alive.

On a barrel-sized dunun drum with a cowbell strapped to the top, she played bellowing bass patterns and tinkling bell tones. I listened, mesmerized, unable to picture myself tapping intricate rhythms with one hand while pounding bass notes with the other. Eventually, I'd learn how. But like everyone else, I got started on the djembe, the West African drum that makes a ruckus in parks.

The simplest rhythms were tricky at first. Mine must have sounded like a galumphing elephant, because a more advanced student beside me would smile and say, "It will come."

Every time I missed a beat, I winced, but the drum teacher kept reminding us to laugh when we goofed up and quickly move on. For the first time in my life, I had a roomful of people showing me with smiles and body language that mistakes were no big deal—and the sounds kept flowing even when I flubbed up. Instead of stressing about my flaws in a therapist's office, I could feel them transforming in a circle of skin-tingling rhythms. Repatterning my perfectionism was all the more powerful because it was happening through music, the very thing that had messed me up in the first place.

We played West African beats layered together in dizzying patterns called polyrhythms, which at first sounded restless and bewildering to me. Imagine someone repeating the word "wa•ter•me•lon" (four beats) while another says "straw•ber•ry" (three beats) and another says "ap•ple" (two beats) all in the same time frame. Drum patterns are still more complex, with sporadic beats left out, unmooring my sense of timing. For weeks, I couldn't get the feel. But as my hands found the patterns, my whole body found the groove. The rhythms stretched and pulled the downbeats in all directions, until my bones started buzzing, down to my toes. I felt electric, soaring on a rhythm high.

At last, I knew what people meant when they called music a euphoric drug.

Tuesday nights became the high point of my week, a circle of warmth and delight. One of the players invited us all to her art exhibition. Another hosted half a dozen of us for a weekend at her island hideaway, where we chopped vegetables in the kitchen, swapped stories, and drummed in her living room far into the night.

THE DRUM CIRCLE gave me experiences worlds apart from anything musical I'd done before. With the cello, playing had always felt like a performance; even alone in a practice room, I felt watched, judged. Still, I couldn't quite put my finger on the difference with drumming until I discovered a verb championed by the late musicologist Christopher Small: "musicking." Music, he argued, is much more than a skill set, a definitive performance, or copyrighted notes on a score. In essence, music is a kind of ritual. It's something humans do—more enjoyably together.

Leaders have always exploited this phenomenon, using anthems to rally the masses and drumbeats to rile up soldiers for war. Marketers have, too. In 1971, one of the most successful advertising campaigns of all time used a jingle to conjure feelings of perfect harmony: "I'd like to buy the world a Coke..."

In the years since this TV ad's hilltop love-in (famously recreated in the hit series *Mad Men*), studies of our brainwaves, behavior, and biochemistry have revealed why we pay big bucks to see rock stars, even though we'll be standing cheek by jowl with sweaty strangers all night. The high doesn't just come from Metallica or the megawatt light show. Drinking in the music with others enhances our delight, intoxicating the gray and white matter between our ears. Whether we register this consciously or not, music draws us irresistibly together.

THE AFRICAN POLYRHYTHMS I learned in drum class come alive when multiple drummers gather to play all the parts. Nobody knows how far back these rhythms go. Drums made of hollowed-out tree trunks tend not to leave an archaeological trace.

But in Germany's Swabian Alps, possible evidence of social bonding through music dates back at least forty thousand years. Archaeologists uncovered flutes made of woolly mammoth and vulture bone alongside stone tools and cooking materials. Nicholas Conard, an American archaeologist at the University of Tübingen, believes these prehistoric flutes weren't just entertainment for hominins huddled in caves, but fixtures in daily life. Music might have been a binding force that encouraged our Ice Age ancestors to stick together, he said, playing a role, along with food and shelter, in our early survival.

From work songs to dance tunes, love ballads to lullabies, songs of the world revolve around social ties. Tight-knit societies weave music into everything they do. In the villages of Uganda, melodies ring out during cooking, hoeing fields, and hanging around the fire. "Music is in every aspect of life in our local communities," said the musician and translator Walusimbi Nsimbambi Haruna, speaking in the documentary film *Throw Down Your Heart*. "If someone lost a relative, their crying is musical."

The film chronicles the American banjo player Béla Fleck's quest through Africa to find the origins of his instrument. In Gambia, he encounters its ancestor: the three-stringed akonting, made from a skin-headed gourd with a stick hammered through it. The akonting shares key features of the American gourd banjos from the 1700s and is played with a nearly identical downstroke style. Despite the banjo's nickname as "America's instrument," it was almost certainly invented in Africa and carried to the New World by slaves.

In Gambian lore, this instrument saved lives. When the first slave ship carried Gambians across the Atlantic, most of them died. But on the second ship, the captives smuggled an akonting on board. The akonting gave them strength, allowing greater numbers on the journey to survive. Then, on American soil, music helped enslaved Africans endure the brutalities of plantation life. Whether or not every detail can be confirmed, this oral history relays an essential truth: music strengthens human ties, in good times and bad, encouraging feelings of hope and unity.

Like family tartans, totems, or crests, songs passed down through generations speak volumes about who we are and where we come from—our affinities, intentions, and loyalties.

Indigenous peoples far and wide are reclaiming dances and drum rhythms to affirm their connections to traditions degraded by colonialism. Jeremy Dutcher, a composer and member of the Tobique First Nation in New Brunswick, Canada, asked Maggie Paul, an Indigenous Elder, what it was like to revive songs her ancestors had sung before. "We brought the music back, we brought the drum back," she said. "Sweat lodges are here. Teepees are going all over the place. Wigwams."

"So you think music had a lot to do with that?" Dutcher said.

"Oh yeah, you got that right," she replied. "Music will bring you back."

The shared meanings and attachments we form through music begin before we can talk. Mothers of every culture sing to their babies. Fathers, too. From our earliest days, melodies and rhythms help us attune to one another.

AROUND THE TIME I started drum lessons, I wrote a song for my child. This wasn't my idea. An older friend, noticing my postpartum struggles, told me a song like this would stay with him forever.

The tune came to me on a rainy afternoon as we did laps around the park. Nestled in the baby carrier, my son wriggled and bleated like a tired lamb, but as I hummed and experimented with different words, he stopped crying. He knew something was up.

I'd been wondering when my lively child would sleep through the night (in the end, it would take two years). Well-meaning people kept saying, "The days are long, but the years are short." *They must have forgotten how long those early days really were.* Even so, thinking of the years ahead gave me a starting point for a song: "Who could believe that one so small / Will grow up to be so tall?"

I pictured my bundle of needs as a little boy who could walk and talk, and thought of all the things we'd do together. "Growing vegetables, eating them too / Swimming in the ocean, building igloos." I imagined my song enveloping him with excitement for the adventures he'd chart for himself. "Maybe you'll be a philosopher / Explore the oceans wide. / Maybe you'll fly to Mars and back, just for the ride."

My song ended with a promise: "Whatever you do, you'll be my child and I love you with all my might..."

The years, I've discovered, really are short. By the time my boy was nine, we had already done most of the things I thought of when he was small. He had gotten tired of hearing his "special song" each night, and I didn't blame him.

Once in a while, though, I still sing it to myself. The words, now bittersweet, remind me of all the moments we've shared and how little time is left before he leaves home, moving somewhere that will no doubt feel as far away as Mars to me. But when the time comes, I plan to celebrate his flight. Writing a song for him helped me figure out what kind of mom I wanted to be. And when I was frazzled, exhausted, or sad, it reminded me of the fierce and enduring love I have for my son. In the end, it comforted me.

This is precisely what the research shows: lullabies soothe mothers as well as their babies, calming the autonomic nervous system and lowering maternal and infant stress.

Tuning in to rhythm and song begins before birth. As early as twenty-seven weeks, an unborn baby may respond to a mother's voice with movements and begin to hear sounds from the outside world. Newborns can remember songs they heard from the watery haven of the womb. In a controlled experiment starting at thirty-five weeks' gestation, mothers played the same melody to their babies twice a day for three weeks, and then stopped. Four weeks after birth, the newborns showed heart-rate responses to the familiar melody not found in babies who didn't hear this music in utero.

Hearing, unlike sight, is fully developed at birth.

Talking to babies helps them feel safe, but singing comforts them more. In one experiment, infants at six to nine months remained calm without their mother in sight for twice as long if they heard singing instead of speech—although both were in Turkish, an unfamiliar language to these babies. In a separate study, visibly distressed infants at eight to ten months smiled more and showed calmer responses on a skin conductance test when a caregiver sang to them instead of talking expressively.

Songs draw us closer in emotional and physiological ways. For one thing, music floods us with oxytocin—aka the "love drug"—a brain chemical and hormone tied to empathy, social connection, and trust.

HUMANS RELEASE OXYTOCIN during cuddling, intimate conversation, and orgasm. Scientists discovered more than a century ago that oxytocin stimulates contractions in childbirth, and milk let-down in breastfeeding. But this charismatic chemical didn't gain sex appeal until the 1990s, when researchers observed that oxytocin enhanced mating rituals in prairie voles—pompom-like rodents that bond for life.

After that, things got weird.

The American "neuroeconomist" Paul Zak raved about oxytocin in a 2011 TED talk that has gained more than 1.8 million views. His tips for a dose of oxytocin: eight hugs a day, or a snort of liquid oxytocin. "You'll be happier," he said, holding up a syringe, "and the world will be a better place." Scrambling to get in on the action, new products such as VeroLabs' Connekt oxytocin spray claimed to help customers "experience more empathy" and "be your true, uninhibited self."

As marketers sang the praises of oxytocin, music researchers joined the chorus. Listening to tunes, they discovered, increased the chemical in cardiac patients recovering in hospital. Those who heard relaxing music for thirty minutes showed surges of oxytocin, while patients without music showed decreasing levels. In separate research, choir members had higher oxytocin after half an hour of singing "California Dreamin'," but not after the same amount of chitchat. Choral singing, the study concluded, encourages the same social bonding found in close friendships and romance.

Based on the hype, you'd think oxytocin was Love Potion no. 9. But this "feel-good" chemical might have gotten credit beyond its due, in studies of marriage as well as music. Shelley E. Taylor, a psychologist at the University of California, Los Angeles, pointed out that oxytocin can surge in women involved in abusive relationships as well as healthy ones. She didn't mince words: "People got carried away with the idea of the cuddle hormone."

Oxytocin has been linked to envy and schadenfreude (pleasure in others' pain). Even as it reinforces social bonds in a tight-knit circle, it may also incite aggression towards other groups. Clearly, this chemical doesn't always bring warm fuzzies. Instead, it may heighten our sensitivity to social cues, both positive and negative.

Oxytocin can also spike in response to stress. So, are soaring levels in a choir a sign of performance anxiety or joyful camaraderie—or both? So far, this tricksy chemical hasn't painted a clear picture of how music strengthens social ties. But there's a lot more to bonding than oxytocin.

AFTER A HIATUS during a work crunch, I made my way back to the drum studio for the first time in months. The thump-thump of the warm-up, a heartbeat rhythm, drew me in. Then we got started on "Djolé," a juicy West African song played for feasts.

The teacher handed me a mallet to beat a solo pattern on the high-pitched kenkeni drum. I felt too rusty to carry the tricky drumbeat myself, so I took cues from the woman beside me playing the big bass dunun. She marked the downbeats with her feet—right, left, right, left—and I began to step in time, just like the chimney sweeps in the *Mary Poppins* song. Soon, my syncopated pattern locked in to hers and as our bodies swayed together, I felt a swell of affection for her, an almost physical love. *Odd*, I thought, snapping out of the reverie. My attraction wasn't sexual—more like sisterly warmth. But I hardly knew her. What was this kinetic attachment that seemed to come out of nowhere?

I found pieces of the puzzle in the work of Laurel Trainor, director of the McMaster Institute for Music and the Mind in Hamilton, Ontario. Trainor's singsong voice reminded me of a flute, her first love before a growing interest in music cognition and psychology led to her current status as a titan in her field.

When we dance or move in synchrony with others, Trainor said, "it turns out that you trust them more—you are more likely to befriend them. You'll cooperate more with them and show more altruistic behaviors towards them."

Despite my rush of affection in drum class, I couldn't wrap my head around this idea until I thought of the close ties that form among rowers, farm workers, synchronized swimmers, and marching armies. Researchers describe it as "muscular bonding."

Whether people are running together, swaying in rocking chairs, or sashaying through simulated dance clubs, study after study has linked synchronous movements to stronger social ties.

As pack animals, we learned to coordinate our movements in game hunts to survive. Rhythmic movements make us feel part of a team, one paper explained, eliciting feel-good emotions that "weaken the psychological boundaries between the self and the group."

Of all the rhythmic things we do, music in particular brings us closer when we move as one. People who drum together experience stronger feelings of group connection than when they drum out of sync. Their heartbeats synchronize, too.

Even before we can talk, bobbing to music with others increases our goodwill. In a series of adorable experiments in Trainor's lab, an assistant wearing a Snugli bounced a baby to music. Facing them, a "stranger" (the experimenter) either bounced with them to the beat or out of sync. Minutes later, on the floor, the baby watched the "stranger" put balls in a bucket or pin dishcloths on a line. Every so often, the experimenter would drop a ball or clothespin to see if the baby would pick it up.

After repeating this experiment with nearly fifty infants, all at fourteen months, the researchers saw a significant difference: babies bounced in sync with the experimenter ended up helping her 50 percent of the time, compared to 35 percent for babies bounced out of sync. A dramatic result after three minutes of music. Rhythmic games such as singing and clapping don't just entertain little ones: they help babies feel closer to us.

As adults, the bonding effects of music tend to fly under our radar. We don't realize that when we see a musician playing, our brainwaves begin to match those of the performer. In a 2020 study, listeners who watched a video of a violinist playing pieces such as "Ave Maria" and "Auld Lang Syne" showed brain synchrony with the violinist, especially in a region that processes rhythmic sounds.

The mirror effect turns up among musicians, too. Guitarists performing together show matching brain activity even if they're playing different parts. This isn't just a case of visual mimicry—monkey see, monkey do. Amateur singers don't need to see each other's faces for their brainwaves to entrain. Through music, noted the study on singers, "two brains make one synchronized mind."

The intimate connection doesn't stop at a handful of people. Music can set the brainwaves of an entire audience oscillating as one. Scientists put this theory to the test at the LIVELab, an eight-million-dollar theater envied by music-cognition experts worldwide. Pioneered by Trainor at McMaster University, the lab looks like a typical concert hall with folding seats and a raised stage, except this hundred-seat venue is equipped with gadgets for monitoring heart rate and brainwaves in musicians as well as spectators.

When audience members in the venue watched a live band, their brainwaves entrained to the music as a group, said Jessica Grahn, co-leader of the preprinted study and an associate professor at Western University's Brain and Mind Institute in London, Ontario. The more their brainwaves synchronized, she said, the greater their pleasure in the performance and experience of group connection.

Most of us have felt this collective energy before, either at a concert or in a hockey arena filled with a thundering chorus of "We will, we will rock you!" We pump our fists and exchange goofy grins, effortlessly communicating with strangers around us. Rhythms and songs speed up the bonding process, dialing us in to a central hub of connection. Music, at a brainwave level, activates our hive mind.

This phenomenon helps explains the popularity of drum circles, campfire sing-alongs, and karaoke clubs, where masses of people with different values and backgrounds come together in a kind of rhythmic gestalt. To quote the neuroscientist Daniel Levitin, "There's this unifying force that comes from the music."

IN A PACKED SPORTS ARENA in downtown Vancouver, the audience was getting restless. More than thirteen thousand people had paid up to sixty bucks to receive pearls of wisdom from the Dalai Lama, spiritual leader of Tibetan Buddhism. But after lining up in the rain, we had waited for nearly an hour inside with nothing to do but stare at the empty stage. His Holiness was late.

Fits of coughing broke out, followed by the odd shout. I began to wonder how long this many people, Buddhist-curious or not, could sit in bum-numbing seats without acting up. Then, out of the darkness, a lone voice rang out: "O Canada." It was so faint, I wasn't sure I'd heard right, but after a second or two, the voice sang again. A quiet fell over the arena as all ears strained to hear the sound. As the voice continued, others joined in, and by the second verse of the national anthem, the entire audience was on its feet singing a welcome song for the monk who would receive honorary Canadian citizenship that day. What a change from the usual loudspeakers prompting a forced display of national pride. This audience had pulled off a spontaneous, jubilant act of song.

When the Dalai Lama finally appeared, bowing in his saffron robes, he smiled, took a seat, and began his talk on cultivating happiness. As always, his warmth and self-deprecating manner won over the room, but as the hordes streamed out of the stadium, I overheard people murmuring about the unexpected happening before the holy man had uttered a word. In those brief moments of song, the audience had demonstrated the values he had flown around the world to teach: mutual understanding, universal love.

As rare as it is for many of us, singing may be fundamental to our biology. When we sing, our inhalations deepen and our lungs exchange more oxygen, calming the parasympathetic nervous system. Singing together strengthens social bonds, especially in large groups. After a choir rehearsal, singers report stronger feelings of happiness, belonging, and social connection—but the effect is most pronounced in groups with two hundred voices or more.

Music expands our capacity for empathy, starting at a young age. In a British school, children aged eight to eleven played musical games for an hour a week, matching each other's melodies and playing together in rhythm. Meanwhile, two groups of similar children either had no games or played games involving drama and storytelling. At the end of the year, all children had tests of empathy, such as identifying and relating to the feelings of a character in a movie clip, but only kids in the music group showed a significant change. Musical games, the study found, double as "empathy education."

Melodies and rhythms help us warm up to each other. And when the world grows dark and everything stops, many of us turn to music for the sense of connection we crave.

ON EASTER SUNDAY, one unforgettable spring, a blind man in a tuxedo strode into the Duomo of Milan, his footsteps echoing across the marbled floor. The colossal pipe organ bellowed as he sang "Ave Maria" to row after row of empty pews. For the first time in his life, Andrea Bocelli, the Italian opera star, performed in front of no one. Instead, he sang for all.

That day, half of humanity—nearly four billion people—was on lockdown to curb the spread of a deadly coronavirus. Huddled at home, many of us had spent weeks watching alarming news and raging debates (to mask or not to mask?) flashing across our screens. But amid the fear and confusion, Bocelli's message from the cathedral rang clear: "Through music, streamed live," he said, "we will hug this wounded Earth's pulsing heart."

At the sound of Bocelli's voice flowing through my laptop, I crumpled in my chair. Something about his warm tenor brought home the enormity of the world's suffering and the horrors still to come. I ached for the sick, the lonely, the doctors and nurses, and most of all, for my parents at the other end of the country. Would I ever see them again?

Operatic singing isn't my thing, but when I clicked on the concert to see what the fuss was about, I found surprising comfort in knowing that more than 40 million people from as far away as Kenya and South Korea had tuned in. One had lost an uncle and a spouse to the virus within days. Bocelli's voice, she wrote on YouTube, was "exactly what I needed to hear." Another gave thanks for the timeless songs and "this feeling of hope and unity." I felt it, too.

Stay-at-home orders went against our primal instinct as social beings: to find safety in numbers in times of threat. Early in the crisis, Italian balcony singers reminded us how to touch each other with song. Strumming their guitars, banging pots and pans, they serenaded the neighbors they'd normally rub shoulders with in bustling piazzas. When Italy was hardest hit, the world cheered at their show of pluck. Later, as the contagion spread, we told ourselves that if the Italians could summon beauty and solidarity in the face of shocking loss, maybe we could, too.

Audiences worldwide found solace in Neil Young's online fireside concerts, house sessions by Alicia Keys, and Yo-Yo Ma's cello videos, posted on social media under #songsofcomfort. Often, though, the songs that touched me most came from voices I'd never heard before, like the Facebook friend who conquered her shyness to post a video of herself singing "You Are My Sunshine."

In times of trouble, "We go back to these roots of [the] innate things that bring us connection and strength," said Sarah Wilson, a clinical neuropsychologist at the University of Melbourne, speaking to the BBC. "What better way to do this than to sing across balconies."

The outpouring of music in the early months of the pandemic reminded me of another global crisis nearly two decades before. In the aftermath of the September 11 attacks, in 2001, I watched one of the world's most cherished musicians pull out all the stops to draw nations together.

EIGHT WEEKS after two passenger jets crashed into the World Trade Center, I boarded an airplane for Syria. Although the War on Terror raged in Afghanistan, nearly two thousand miles to the east, few American journalists felt safe enough to accept an invitation to the Aga Khan Award for Architecture, held that year in Aleppo. But I was childless and thirty-one. I jumped at the chance to tour this ancient city, continuously inhabited for at least five thousand years, and wander through souks brimming with silks, tapestries, and fragrant cubes of laurel and olive oil soap. (A dozen years after my visit, Aleppo and its warmhearted people were devastated by civil war; see notes.)

In the gleaming hotel where I was staying, courtesy of the Aga Khan, leader of the Ismaili Muslims, I waited by the elevator to go up to my room. When the doors opened, I gasped. There was Yo-Yo Ma. Even without his cello, I would recognize him anywhere, but I doubted he'd remember me from backstage when I was sixteen. As we traded places, I smiled without asking the question on my lips: *What are you doing here?*

That night at the awards ceremony, I learned that Ma had flown to the Islamic world to bridge cultural divides.

The evening started with a military parade outside Aleppo's towering citadel, where spotlights illuminated the crumbling fortifications at the top. Along with five hundred international guests, I climbed the massive stone steps leading to the castle, crossing a bridge over a moat supported by eight enormous arches. We feasted in the dining hall, where waiters in spotless uniforms carried table-sized platters of tender lamb studded with pistachios and pomegranate seeds. Afterwards, Syrian hosts ushered everyone into a fifteenth-century throne room. I gazed at the ceiling of inlaid wood, as intricate as a Persian tapestry. Then I listened as Ma and an East-West mélange of musicians played variations on a theme of cultural exchange.

The cellist had teamed up with the ethnomusicologist Theodore Levin to create the Silkroad Ensemble, named for the historic trade route that had linked the goods and cultures of Asia and Central Asia with those of the West for more than a thousand years. Ma, said Levin, regarded the Silk Road as "a metaphor for human connectedness, creativity, and imagination." That night in Aleppo marked the group's first performance on the Silk Road itself. But as American troops descended upon neighboring Islamic countries, how would they be received?

I could imagine their jitters. The audience included dignitaries from Iraq and Iran, nations soon to be named part of U.S. President George W. Bush's "axis of evil." I admired the musicians' poise as they played for the assembly of Arabic, Kurdish, Hindi, English, and Farsi speakers. And now, through the lens of neuroscience, I can picture the brainwaves of five hundred guests—many from warring countries—synchronizing to the rhythms of Ma's velvety cello mingling with instruments from Asian, Southeast Asian, and Central Asian lands.

To my ears, the most exquisite music that evening was a Silkroad commission by Iranian composer Kayhan Kalhor entitled "Blue as the Turquoise Night of Neyshabur." Ma's cello struck a meditative note, joined by violins, a kemanche (spiked fiddle), santur (hammered dulcimer), and ney (bamboo flute). As the graceful rhythms infused the hall, I could almost picture the inky sky above the Silk Road city of Neyshabur during the "fourth time of day," when deepest night leads to dawn.

From what I could tell in the throne room, the enthusiasm for Yo-Yo Ma's group went beyond polite applause. An Aleppo-born architect, Nasser Rabbat, told me he enjoyed the East-West fusion more than either traditional Central Asian or Western classical music on its own.

Nine months later, I had the opportunity to interview Ma just before the Silkroad Ensemble's first performance in Vancouver. Ma told me the project had reinvigorated his approach to the standard cello repertoire, "because the work with the percussionists has shown me how many different kinds of grooves there are in rhythm. So whether I'm playing Haydn or Tchaikovsky, I'm looking for that," he said. "A groove is not culture-centric."

He added that cultural exchange works both ways, and relies on empathy and trust: "I've always started from the premise that if I know what is precious to you, and you know what is precious to me, as we meet—as strangers, even—we can start a very different dialogue."

Music helps us reach each other beyond the minefield of words. In times of conflict and peace, it may be one of the strongest social glues we have. As I've learned from my own family, though, sometimes this "glue" takes extra effort to make it stick.

MY HUSBAND has a soft spot for Disney movie songs, which bring back fond memories of watching *The Wonderful World of Disney* with his family as a child. Sadly, I can't stand the greatest hits of *Aladdin* or *The Little Mermaid.*

I like Congolese soukous beats. Scott finds them chaotic. But that's how I feel about the top-forty radio he switches on when he gets behind the wheel. One day I mentioned that David Byrne calls top-forty the "fast food" of music. Scott retorted that he didn't care what David Byrne thinks—he's never been a Talking Heads fan anyway. My husband likes what he likes and has no patience for musical snobbery. I admire his attitude. Compared to me, Scott tends to be more open to new sounds. He's the one with the music-recognition app, Shazamming the latest songs in coffee shops.

More than once, though, I've hurt his feelings by groaning about a song he loves, and he's retaliated by vetoing some of mine. Bonding over music hasn't come easily for us. Given my intense relationship to music, this irony hasn't escaped me. But since I wouldn't trade Scott for all the songs in the world, I've learned to treat this slight mismatch as a "growth area."

The research on social bonding has given us strategies for hacking a stronger musical connection. On a road trip, we'll listen to scores of albums, searching for new tracks to add to our "Scott + Adriana" playlist (his idea!). Using this method, we've discovered a shared taste for vintage Italian movie soundtracks, California surf ska, and bands like Thievery Corporation and Arcade Fire. When we put in the time to find common musical ground, brainwaves and neurochemicals usually take care of the rest.

Luckily for me, Scott has never bought into the white-guys-can't-dance myth. Even so, enrolling in a salsa class together felt risky because I had already learned Latin rhythms in drum class. Scott caught on fast, though, having traveled in Central America, and as the dance steps got trickier, they challenged me, too. Often, we headed to salsa class dragging our feet, but after an hour of practicing moves such as enchufla and dile que no, we'd walk home laughing and holding hands, revved by the spicy rhythms.

Next up, our son. By the age of ten, he was as repulsed by my world music as I am by his Minecraft-themed pop. Mutual hatred of another generation's music is practically a rite of passage, so I wouldn't have fretted if new research hadn't given me pause. In a University of Arizona study, college students who shared music with their parents in childhood, especially in their teen years, reported closer relationships with their parents later on. Skeptical, I assumed families who listened to music together had a stronger relationship all along. Yet compared to playing board games or eating meals together, in this study, music strengthened family bonds most. Another eye-opener: casual activities, such as listening to playlists in the car, had as much or more impact as making music with parents in a band or quartet.

For me, this came as a relief because, much to my chagrin, I hadn't succeeded in creating a family culture around music-making any more than my parents had. In my perfect world, we'd play instruments together, but my son is more into computer animation and my husband would rather go mountain biking or skiing.

Technology has made it easier for us to connect through music. One winter our family had a blast watching reruns of the high-school drama *Glee*. For the first four seasons (until the show lost the plot), we spent evenings belting out Queen's "Bohemian Rhapsody" and Journey's "Don't Stop Believin'." Later, we watched *Soul*, the Pixar animation about a school band teacher, then *Coda*, a drama about a girl raised in a deaf family who discovers she loves to sing. Bonding with my family over tuneful TV was more fun than I thought.

FIVE YEARS AFTER I took up hand drumming, my teacher lost her rehearsal space and moved our sessions to her home in a basement suite. Our numbers dwindled from about fifteen regulars to seven. Our sound shrank, too. An irate neighbor knocked on the door each time our playing reached the volume of a loud TV. To placate him, we lightened our drumming to a pitter-patter and used chopsticks to tap out the bell patterns, barely striking the metal. My teacher made the most of these constraints. "Softer playing can help us focus on the sound quality of our drumming." She encouraged us to sink into the feeling of the most basic pulse: thump, thump, thump. Over time, we became less of a drum circle and more of a women's group, sharing the details of our lives over herbal tea. Much as I cherished my connection with them, I missed the electrifying rhythms we'd played before. Drum lessons no longer left me buzzing inside.

With mixed feelings and gratitude for the teacher who had taught me so much, I drifted away from the group.

I had my eye on a samba band who wore silly costumes and rainbow-colored wigs at Vancouver festivals and Pride parades, with bright ropes of LED lights strung around their metal drums. The group, Bloco Energia, took their name from the street bands who march in blocos (blocks) during Brazilian Carnival. I wanted a dose of their zany energy. And so, fifteen years after rambling through the streets of Recife with Kleber, I finally dove into the rhythms of Brazil.

The group met weekly in a windowless rehearsal studio that got stuffier with every drumroll. A flaccid sofa at one end faced a collection of metal drums stacked on the floor, some the size of side tables. Duffel bags held an assortment of high-pitched percussion instruments that cut through the rat-a-tat of the snare drums and the booming bass of the large surdo drums. People took turns with conical agogô bells; cylindrical rattles called ganzás, played with a chuga-chuga swing, like a choo-choo train; and tiny hand-held tamborim drums struck with a bundle of nylon sticks that made an ear-splitting *crack*!

Earplugs, anyone? The band kept a box of them handy, because no one could stand playing without a pair.

I learned not to expect step-by-step instructions or meditations on the fullness of each pulse. Instead, one of the leaders handed me a drum with a pair of mallets and teamed me up with a seasoned player who knew all the moves. Beginners like me had to play along as best we could, sink or swim. As the rhythms flew by at warp speed, the fake-it-till-you-make-it approach made it impossible for me to fixate on every missed beat. And mistakes were hard to hear amid the roar of metal rattles, whistle calls, and crackling snare drums.

Samba gave me a jolt of adrenaline. Instead of softly drumming with women in midlife, I worked up a sweat next to musicians of all ages and genders, and a dozen nationalities. This was festive music—extroverted, in your face. "The energy of the crowd kind of feeds us," said one of the band leaders, Marco Castrucci, "and we feed them."

The high point for me came the day we played at a spring fundraiser for my son's school. More than a dozen of us showed up in matching white jeans and frog-green T-shirts plastered with the Bloco Energia logo. Lined up in the schoolyard, metal drums dangling from our waists, we danced side to side and front to back, timing our choreography to the cacophony of our drums. When the samba whistle blew, we launched into a sequence of rhythmic stunts, clacking our mallets together, twirling them in the air, and yelling “Hey!” or “Ha!” together in time. This must be how marching armies felt—united, strong, unstoppable.

The wall of samba drowned out the traffic sounds and gleeful screams of children in the bouncy castle. Whiffs of pizza and popcorn drifted our way as the crowd of families munched and watched. I remembered to smile as I pounded my drum, even if I forgot a riff and had to air-drum for a second or two. For once, I didn’t care what the other mothers at the school thought. I wasn’t a baker, a candy-floss maker, or a dunk-tank volunteer. Upbeat music was the best thing I could offer, and there in the crowd, my son was cheering me with pride. I was a real musician again, bringing rhythm and song to life.

That night, I spotted pink marks on my legs from the metal drum banging against my knees. They made me smile. Samba battle scars. Playing in the group was brash and fun. But blocos are aptly named: a samba band is nothing less than a deafening block of sound. After a while, I began to miss the intimacy of softer instruments and the chance to play them at home.

Once again, I abandoned one music group in search of another. But this time, I had no regrets. Instead of committing to one instrument for life, I could play the field, try something for a while and come back to it later, or not.

I’d gone from seventeen years of cello to seven years of hand drumming to just a year of samba. The rhythm of my musical life was speeding up.

—9—
The Beat Goes On

NEVER DID I YEARN to play a ukulele. The plinky-plonky tones turned me off, along with the hipster fad of strumming Radiohead and David Bowie songs on the Hawaiian four-string. Maybe I couldn't get past the sound of Tiny Tim singing "Tiptoe Through the Tulips" in his warbling falsetto. Talk about an earworm.

Then there were the ukulele jokes.

After my parents' move to a historic fishing village near Liverpool, Nova Scotia, my stepfather, Russell, volunteered as an organizer for the first International Ukulele Ceilidh—that's Celtic music with a hula vibe. Although he'd never played the breezy instrument himself, he celebrated the occasion by publishing a book of ukulele zingers along the lines of "What do you tell a beginner on the ukulele?"

"Uke can do it."

Every time I visited, Russell told the same ukulele jokes. I would groan, begging him to stop, but he'd keep going, chuckling with glee.

Eventually, the joke was on me. In my late forties, I found myself in a music shop buying a copy of *Jumpin' Jim's Ukulele Island: 31 Tropical Tunes Arranged for Uke*. For once in my life,

I wanted to learn songs that friends and family might enjoy around a campfire. African drums didn't fit the bill and guitar would be an iffy choice for my tendinitis-prone hands. That left the ukulele—small, cheap, and easy to play. But I couldn't seem to unzip the ukulele case without getting bogged down in midlife doubts. After all, how much time was I willing to put into learning "Limbo Rock"? Could I really see myself leading a campfire sing-along? *Who was I kidding?*

After a certain age, trying something new dredges up a mess of identity issues and regrets, combined with a gnawing sense of time slipping away. But: If not now, when? One morning, this argument won the day. After leafing through *Jumpin' Jim's* arrangements, I tuned the ukulele using an online app and gave Bob Marley's "Three Little Birds" a go.

At first I sounded more like a crow, rasping along to simple chords my fingers struggled to play. ("Don't worry," plink-plonk—*damn!*—"about a thing...") People assume that having a music background makes learning a new instrument a cinch. Not quite. Yes, I could read music, but my fingers still had to figure out how to scrunch themselves between the frets. Singing while playing didn't come naturally either. I had to work out the rhythm of the melody separately from the chords and then fit them together. It took a few days to get the hang of it ("Uke can do it"). Next, I tried strumming on the off-beat to add a reggae feel. Better, but still nothing close to the Bob Marley playing in my head.

Don't worry, I reminded myself. At this point, my days as a cellist with ten thousand hours under my belt amounted to a past life. I had little hope of mastering any instrument, not even the ukulele, butt of jokes. So, I could either make peace with playing badly or deprive myself of making music that lightened me up. Each song I learned on the ukulele brought me closer to the laid-back musician I wanted to be. As Louis Prima crooned, it's later than we think.

Soon I would be fifty. The inevitability of aging—and let's face it, dying—had became a preoccupation of mine. And not just for my sake.

MY MOTHER INSISTS she's always been scatterbrained. When she was eleven years old, her classmates in boarding school nicknamed her "Granny" (a pet name she wore with pride).

As a child, I had trouble holding Mom's attention because the flit of a butterfly or a wash of sunlight on wildflowers could easily distract her. "Look! Isn't it glorious?" she'd exclaim. I grew up admiring her artist's brain. But in her early seventies, she became more absentminded. Spacey, even. I could feel her slipping away.

When we talked over the phone, she kept saying she didn't know what day it was. "Just turn on your computer, Mom." We had the same conversation over and over. Then she lost her purse, along with her credit cards and driver's license. "Mom, it's illegal to drive without a license," I'd remind her, urging her to call the motor vehicle bureau. "Oh yes," she'd reply, "I suppose I should."

She ended up driving for three months without a license until the car broke down. No one was hurt, but this couldn't go on. Mom was the only caregiver for my stepfather, who was diagnosed with Alzheimer's in his early seventies. Russell had a family history of the disease, along with another risk factor—partial deafness—likely from his habit of playing with explosives as a child after the Second World War. (He got tips on making gunpowder from a Victorian how-to manual, *Every Boy's Book*.)

My siblings and I tried to arrange outside care for our parents in their quaint 1832 house, but Mom kept turning the aides away. When a doctor found an angry patch of malignant melanoma on her shoulder, we talked them into a temporary move to an assisted living facility. "This way, Russell will be safe while you're having surgery, Mom." But I had a feeling they'd never get back home.

Watching my mother declining in her early seventies gutted me. Dementia doesn't run in our family; my grandmother had stayed sharp until her early nineties. But Mom, the health nut, had refused to take medication for two decades of runaway blood pressure. This might explain the CT scan showing vascular damage in her brain. Regardless, in my eyes, she was too young to be losing her faculties. Friends her age were still painting in their studios—and making art was Mom's greatest joy.

Just twenty-four years separated us, a gap that made me worry about my own cognitive future, too. So far, I had no memory problems or health issues, but how long would that last?

As usual, I hit the books. Combing through the latest studies on aging, I discovered a growing body of evidence showing that generous doses of music can enhance our quality of life, and maybe even preserve some of our faculties until the end of our days.

MUSIC TRAINING, like second-language skills, might offer some protection against cognitive decline, although it's too soon to say for sure.

In a series of studies, the more years people had played an instrument by the age of sixty and up, the higher their scores on tests of working memory and executive functioning—the type of thinking needed to carry out everyday tasks such as managing a bank account or keeping track of doctor's appointments. The link between music and mental faculties persisted regardless of education level, physical activity, and other lifestyle factors. In a separate study of adults over age fifty-five, stronger auditory processing tied to music training lingered even in those who hadn't touched their instrument in forty years. (Good news for anyone who gritted their teeth through music lessons as a kid.)

Before digging in further, though, I reminded myself to take these findings with a bucket of salt. Dementia is an umbrella term for Alzheimer's and a cluster of other diseases that alter memory, mood, cognition, and behavior. Scientists know too little about these diseases and their insidious causes to declare that music, or any other single activity, can prevent dementia.

Not surprisingly, a 2020 review emphasized that more research is needed to confirm whether musical activities can help curb the progression of aging and dementia in the brain. Nevertheless, the same review described music as a "promising cognitive intervention" for older adults. The working hypothesis: music makes cognitive demands that promote neuroplasticity—the brain's ability to change itself—in key regions involved in learning, thinking, and memory.

Some of the most persuasive evidence I've seen comes from a small study of Swedish twins. The elderly twins, mostly non-identical, grew up in the same family environment (a major influence on long-term health) and shared at least half of any genes that might affect their dementia risk. None was a professional musician, yet those who played an instrument were much less likely to develop dementia, or mild cognitive impairment, than their non-musician twin. (To rule out "reverse causation" from cognitive problems that might have forced a musician to stop playing, the study defined musicians as those who had either played regularly or occasionally up until at least five years before a cognitive assessment.) The researchers didn't know when the musician siblings took up an instrument, but the results suggest "playing an instrument has a positive influence on neuroplasticity, regardless of what age one begins."

Wishful thinking? Not necessarily. Latecomers to music were put to the test in a Florida study offering piano lessons to a group of non-musicians aged sixty to eighty-five. All agreed to practice at home for at least three hours a week on top of a weekly private lesson. After six months, the beginner pianists performed significantly better on tests of attention and working memory than those who didn't take music lessons.

Studies like this bolster the idea that music training, even later in life, may offer "transfer effects," meaning skills that spill over into other activities. But so far, none of the research is conclusive. The age-related benefits tied to musical activities might have some other explanation, just like the disproven link between music training and IQ (see chapter 5). And even if music does have a protective effect, we still don't know if music supports healthy aging any better than, say, learning a new language or taking up a new sport.

In general, we get the best neurological workout by pushing our limits. The neuroscientist Daniel Levitin, whose most recent book is *Successful Aging*, spells it out: "If you're just thinking the same thoughts and doing the same things you always did, it doesn't force the brain to keep making new pathways available to it." In other words, a pianist who keeps practicing the same sonatas, or a banjo player strumming the same bluegrass, might want to mix it up. (On the bright side, people like me can reframe our musical dabbling as "neuroplasticity building" instead of "mistress of none.")

One last caveat: with music, as with a fine bottle of Bordeaux, there can be too much of a good thing. In an eye-opening study, scientists used MRI to detect age-related deterioration—brain shrinkage—linked to Alzheimer's and other dementias.

Compared to non-musicians, instrumentalists had younger-looking brains. However, the amateur players had plumper lobes than professional musicians, who might have led less well-rounded lives. Brain plasticity may be "maladaptive," the researchers wrote, "when it is driven by extensively performed and highly specialized repetitive sensorimotor activities for many hours per day, as well as during stressful public performance periods." (Fingers crossed I quit the cello in time.)

If our brains could talk, they might tell aspiring musicians to stop trying so hard and embrace the perks of being an amateur. In English, we say "amateur" to describe someone who is unpaid, unprofessional, incompetent, or inept (a "rank amateur"). Unlike the French and Italians, we have forgotten the word's Latin root: amare. To love.

LEGIONS OF GEN-XERS want to take up electric bass, kit drumming, or guitar before they die. On a bucket-list survey at Ranker.com, learning to play a musical instrument was tenth on the list among voters born between 1965 and 1980—the highest of any goal requiring a new skill set. In contrast, learning a new language took thirty-eighth place, meditation was in forty-fourth place, and writing a novel was near the bottom, in the seventy-ninth spot.

The question is, how many would-be musicians will ever get around to their bucket list?

Adults often ignore the urge to learn an instrument for fear of playing the wrong notes, says Levitin, or because of times they felt ridiculed in school music classes. Yet the musically curious keep asking him for tips on how to get started. His answer: "Choose the right instrument." If you're in midlife and your heart is set on violin, "you're inviting all kinds of frustration." Coaxing a beautiful note from a violin can take years, he warns, and until that point, violin playing can sound "like somebody strangling a cat."

Others long to play guitar because it's portable and fun to strum at parties, but Levitin, a guitarist himself, emphasizes how much time it takes to develop the calluses and fine-motor coordination needed to play any stringed instrument. "I say, don't make it hard on yourself." Instead, he recommends slapping down eighty bucks to buy a cheap electronic keyboard that anyone can play in tune, and "getting to the part that's fun."

A solid strategy. On the other hand, why settle?

I chose the ukulele for practical reasons, but it's not my first instrument (nor will it be my last). I'll bet most beginners are more likely to get past the learning curve on an instrument they truly love. If playing viola is on your bucket list, plinking away on a tinny keyboard might seem a poor consolation, like visiting Disney's Eiffel Tower in Florida when you dream of Paris.

Miranda Wilson, an associate professor of cello at the University of Idaho, insists there's no age limit for learning to play the cello, viola, or violin. Writing in *Strings* magazine, she quoted Frank Nichols, age sixty-eight, who took up the cello in retirement after deciding what he wanted to do "when I grew up." Sixty-year-old Carol Haynes gave in to the same yearning: "The day I started playing the cello, my life changed utterly in ways that I could not have imagined. In many ways, the anniversary of my first lesson is more significant than my birthday."

Mustering the courage for the first lesson might be the hardest part. Wilson recommends finding a teacher who enjoys working with adults, and setting realistic goals (maybe orchestral playing, not solo recitals). A mature student might have years of poor posture and muscle memory to unlearn; on the upside, adults often understand the logic behind musical notation quicker than children do, and tend to have stronger motivation to practice.

Peter Kennedy, a business journalist and former colleague of mine, makes another case for picking a favorite instrument. "There's never been a better time to learn guitar," he said, noting the gold mine of free instructional videos on YouTube.

Peter speaks with a lilt from his boyhood in Northern Ireland. He had two years of piano lessons before the age of eleven, and then gave up music for five decades, "which I kind of regret now." He bought himself a guitar at age sixty-one, but he had it for at least a month before he finally took it out of the box. Nine months later, using a songbook and online tutorials, he had taught himself classics ranging from "Dead Flowers" by the Rolling Stones to Neil Diamond's "Sweet Caroline." Now he's hooked. "It's just amazing how easy some of these songs are to play," he said. "You know, The Beatles' 'Love Me Do' is really just G major to a C major chord."

Peter sees the guitar as a way to keep challenging himself when retires, "because I'm not going to be playing soccer anymore, right?" I asked him how he gets himself to practice. Working from home, he keeps his guitar on a stand beside his desk "so if I have free time, I'll just sort of pluck away on it. It's never out of sight, out of mind." Excellent advice.

I ADMIRE PETER'S forward thinking, because senescence scares me. When I picture myself in old age, my biggest fear—other than the risk of losing my faculties—is loneliness.

As a child, I often felt alone, cut off from broader webs of family and community. We lived nearly a thousand miles away from my mother's smattering of relatives in Nova Scotia. My stepfather's family lived in England, and the few relatives on my biological father's side remained in Poland behind the Iron Curtain. Adding to the distance, my parents moved to Nova Scotia shortly after I settled in Vancouver, a full day's journey and four time zones away.

Vancouver has a reputation as a hub full of transient and standoffish people, a cliché that's partly true. In this city of glass towers, people come and go. While I cherish the social connections I have now, will they be there later on? I dread the thought of my husband dying first, leaving me on my own.

Loneliness in America has been declared a public health crisis, carrying health risks equivalent to smoking fifteen cigarettes a day. Social isolation makes people feel empty and unwanted, even if others are around. In a landmark study of American retirees, 43 percent reported loneliness, though just 18 percent lived alone. Compared to those with healthy social lives, the lonely ones had a 45-percent increased risk of dying within six years.

This made me shudder: loneliness leaves its mark on the brain. In the largest-ever brain-imaging study of lonely people (based on self-reports), a team at the Montreal Neurological Institute discovered thicker tissues in the default network, the part we use to talk to ourselves and communicate with other brain regions. Researchers speculated that those who lack meaningful social connections may compensate through daydreaming and a kind of inner dialogue. While this kind of thinking might sound adaptive, it's not. Chronic loneliness increases the risk of dementia by 50 percent. A cruel double whammy.

Fortunately, if I ever find myself achingly alone, I'll know what to do: join a choir. As Pollyanna as this might sound, singing with others offers a fast-acting antidote to loneliness. In a large California study funded by the U.S. National Institutes of Health, 390 adults around age seventy sang in a choir for six months.

Compared to those on a wait list, the beginner singers reported a dramatic drop in loneliness and renewed interest in life. The experimental choir, Community of Voices, showed such strong benefits in people from diverse backgrounds that the organizers produced a free manual to encourage similar choirs to flourish.

While any hobby that draws like-minded souls can help people feel less isolated, group singing may quell loneliness faster. Adults in a University of Oxford study who took courses in creative writing or crafting gradually got to know each other after seven months. But those who enrolled in singing felt closer within the first month. Compared to other activities, "singing broke the ice better." (For more on social bonding, see chapter 8.)

Unlike many activities people enjoy in retirement, from international travel to golf, singing in a choir doesn't require disposable income or physical agility. The downside: those who see themselves as unmusical might rather die of loneliness than give it a try.

ASKED TO SING the national anthem, many will laugh and say, "Cover your ears." Humor like this has a beat-you-to-it effect: others are less likely to mock someone's singing if they've already mocked it themselves. Very few "bad singers," though, are actually tone deaf (more accurately called amusia; see chapter 3). One way to self-diagnose is to record yourself singing: if you can tell you're singing out of tune, you probably don't have amusia.

Many "tin-eared" adults grew up with parents who didn't sing to them because of their own hang-ups. As one generation passes their singing fears to the next, tone deafness becomes a family truism. Family attitudes about music, combined with musical activities at home, predict with 74-percent accuracy which children will choose to make music, regardless of their own ability.

"Our family doesn't do music," an athletic friend told me, insisting that her voice is so bad that her housemates have asked her not to sing around them, "so I sing in the car or shower." Her story made me sad because, with rare exceptions, even severely pitch-challenged people can learn to sing. Most just need help training the ear and developing vocal control, along with confidence. William A. Mathieu, a California-based vocal teacher and composer, starts by showing adult students how to mimic sustained sounds, such as a siren, followed by intermittent tones. Then he'll ask a volunteer to sing any note they can manage, and he'll match it. For the first time in their life, he told the BBC, the student gets a taste of what it's like to sing in tune with someone else. Within eight to ten classes, most students are ready to sing different parts as a group.

Rehab for "bad singers" has taken off in England as well. Shy warblers can join one of several "can't sing" choirs, or take a class at London's Morley College called "Tone Deaf? No Way."

For those terrified of singing, a life without crooning might not sound so bad. After all, not everyone is cut out to be Taylor Swift or Kanye West. But adults who don't sing deprive themselves of an age-old source of emotional well-being. In people coping with depression, Parkinson's, or stroke, group singing has been linked to better moods and quality of life.

Galya Chatterton, a certified clinical counselor in Vancouver, told me she sometimes prescribes singing to patients who have suffered traumatic experiences that did not allow them to express themselves at the time. One way to heal from these experiences, she said, is to exercise our own voice and let it be heard.

Our voice is the only instrument that comes from within us. At a fundamental level, it communicates who we are. If we haven't sung by the time we're fifty, sixty, or seventy, we probably have a lifetime of good reasons. But again: If not now, when?

IN MY LATE FORTIES, it occurred to me that even if my hands stiffen down the road, with any luck, I'll always have a voice. Mine is not especially beautiful, but compared to the pitch-challenged, I had it easy. Years of solfège (a system for sight-reading and ear-training) had taught me to sing in tune. On the downside, off-key singing rattles me, so I couldn't see myself joining a choir open to all. The alternative: audition. *But I'd sworn, never again!*

Searching for choirs online, I zeroed in on one that performed music from all corners of the globe. "Somewhere along the way," wrote Karla Mundy, director of the Rhythm 'N' Roots choir, "many of us became musical spectators instead of participants. Singing is a joy that belongs to all of us." *Bingo.*

I prepared for my audition during one of my last visits with my parents before their move into a care home. On walks by the shore, I sang along to Eva Cassidy's cover of Sting's "Fields of Gold." The ballad, tinged with sadness, speaks of love, death, summer days, and memories. As I learned the lyrics, I thought about my parents' future, and sang for my mom.

Back in Vancouver, my audition took place in Mundy's cozy dining room strewn with toys. My pulse raced as much as it ever had on stage (some things never change). I sprinted through the song and didn't nail every note, but Karla's smile kept me going. Then she asked me to memorize a tune she played on the piano and keep singing while she harmonized on top. After my years of orchestra training, holding my part was a cinch.

"Welcome to the choir!" she said. I beamed.

At my first rehearsal, most of the singers were at least a decade older than me. My spirits sagged. I had hoped to meet peers, maybe future friends. *Don't be so ageist*, I scolded myself. Soon enough, I had to buckle down to keep up.

Karla led rehearsals at a breakneck pace, covering a dozen or more songs each night. Without being a stickler, she got results. "Sopranos," she'd say, "can you make it sound just a bit lighter, like you're making popcorn?" Indeed, they could.

The choir performed songs in their original languages, in five-, six-, or seven-part harmony—all from memory. The group included a professional linguist who gave tips on pronunciation in Croatian, Gaelic, Bulgarian, Estonian, and Portuguese. A retired doctor in his seventies told me he had joined the group to keep his mind sharp. Like him, none of these singers seemed to have any trouble remembering their parts. The contrast with my parents' state gave me a pang; still, I liked getting to know adults their age with such impressive memory skills and musical chops. When we performed by heart, their faces shone.

As I learned about breath control, my body became like a cello, resonating from the inside. I had felt the reverberations of music many times before, but this was a different sensation, a kind of internal massage. An outer one, too, as the soundwaves of sixty voices—thick, deep, smoky, and nectar sweet—pulsed through my body and mine through theirs, dissolving the lines between skin and song.

I WAS LUCKY ENOUGH to have a muse for taking up singing in midlife. Madeleine Pouliot, a family friend, was fifty years old when she joined Ottawa's Big Soul Project. She couldn't read music, but that didn't matter one bit. Four years later, she and her choir mates shared the stage with Barbra Streisand in the diva's Back to Brooklyn show. The finale called for a massive choir and Big Soul, an open-to-everyone group, was the biggest in town. Madeleine told me she'll never forget singing in the alto section before a live orchestra in a stadium filled with ten thousand fans cheering for "Babs," who dazzled in a persimmon-colored dress. Big Soul's performance might not have been flawless, but "every single person in our choir sings from their heart—and that's what people respond to."

Over time, Madeleine learned to belt out solos "like friggin' Beyoncé, in my mind, at least." Her flamboyant style caught the eye of an *Ottawa Citizen* reporter, who profiled her in a feature entitled "The Chorister." She told him the choir lifted her spirits no matter what was happening in her life. "It's what keeps me sane. I have to be a step from dead before I won't show up here on a Monday."

Six years later, that day came.

In the spring of 2017, she underwent the first of two surgeries to remove a grapefruit-sized tumor in her brain. The choir became a lifeline for Madeleine, who lived alone at the time. Singers filled her hospital room with flowers, planted seeds in her garden, and passed around a hat to have meals delivered to her home.

Now fully recovered, Madeleine volunteers as assistant choir director, helping the group prepare for concerts to raise funds for causes ranging from the local food bank to a microcredit program for women in Tanzania. Community building is the choir's raison d'être, she said: "Almost everybody in here will tell you that this is like their separate family."

Stories like Madeleine's remind me that healthy aging doesn't mean we'll live out our days unscathed. Instead, it's a decision to cultivate joy and connection with others, regardless of the indignities our minds and bodies may face.

MUCH AS I FEAR DEMENTIA, it's not the fastest-growing neurological disorder. That grim title goes to Parkinson's disease. Although dementia remains far more common, the number of people with Parkinson's has more than doubled worldwide over the past thirty years.

It starts with a trembling hand, slurred speech, or arms that no longer swing when someone walks. Over time, people with Parkinson's often struggle to stay steady and may eventually lose bladder control.

I have no inkling of what my risk might be, as the causes of this incurable disease remain unclear. Scientists do know that Parkinson's involves a drop in dopamine, the brain chemical that not only prods us to seek pleasure, but also helps us move quickly and smoothly. Fortunately, if my husband or I ever develop Parkinson's, our salsa moves might help us stay on our toes. Jessica Grahn, director of the Music and Neuroscience Lab at Western University in London, Ontario, reveals a bright spot: "Music has something that revs up our motor function—rhythm."

When people with Parkinson's dance to music, said Meg Morris, professor of clinical rehabilitation at La Trobe University in Melbourne, Australia, their movements tend to speed up and they show less freezing of gait. "What the music does is to trigger the movements trapped inside."

In other words, music provides an external rhythm to compensate for brain areas that have gone off beat.

At the University of Colorado, neurologists are studying whether playing musical instruments can help restore fluid hand and finger movements in Parkinson's. Working with a music therapist, participants start with rhythmic hand and finger exercises and then progress to small clackers called castanets, followed by piano keyboards.

Meanwhile, the case for dancing has grown by leaps and bounds. When a Toronto team followed people with mild to moderate Parkinson's for three and a half years, they found unmistakable improvements in those who danced. Participants—mostly men who had seldom danced before—took a ninety-minute dance class to live music once a week. Compared to a group that didn't have the lessons, the dancers had fewer speech problems and tremors and less muscle rigidity. Their moods improved, too.

Commenting on the 2021 study in the *Toronto Star*, Karolina Bearss, a neuroscience researcher at York University, said the normal progression of the disease "is not seen in our dancers with Parkinson's." Dancing recruits our senses of seeing, hearing, touch, and positioning in space, while also offering a social outlet. Other exercise doesn't cover all these bases, she said: "There's so much more to dance."

Dancing isn't a panacea, though. After a few weeks without moving to music, the benefits wear off. But compared to weightlifting or stationary bike sessions, dance classes tend to keep people coming back.

Patricia Needle joined a class called Dance for Parkinson's shortly after her diagnosis at age sixty-one. Gazing around the room, she noticed some people with late-stage Parkinson's looking hunched and wobbly. But as the music played and the teacher taught the first move, "I got it," Needle said. "We all got it." Similar classes have run since 2001, when a Brooklyn nonprofit approached the Mark Morris Dance Group to see if their modern dancers would teach classes to adults with the disease. Since then, the Dance for Parkinson's model has flourished in more than twenty-five countries.

A year after joining the class, Needle participated in a dance performance in San Francisco, interpreting waves of kelp moving underwater. While the supple moves of the troupe's able-bodied dancers contrasted to her stiffer ones, "I was deliriously happy." Needle, who had worked for thirty-three years as a registered nurse, said she was "awed by the power of dance to transform and alleviate pain."

Her words will stay with me if I ever find myself in her shoes.

MANY OF US have seen videos of an elderly man or woman seated in a wheelchair with an unchanging gaze. At the sound of a familiar tune, their eyes light up and hands move to the beat.

Of all the clips like these, I'll never forget the footage of a former prima ballerina transported by the music of Tchaikovsky's *Swan Lake*. Silver-haired and painfully thin, Marta González Saldaña had Alzheimer's and looked to be in her eighties. At first, she barely waved a hand to the music, her face downcast, but at the blast of the horns, she peered around the room as if onstage, her arms fluttering to Tchaikovsky in perfect time. As the pounding brass drew the music to its crushing finale, she enfolded herself in her frail, wing-like arms. She was the dying swan.

The elderly dancer showed what music meant to her with every scintilla of her being. Although dementia had robbed her of coherent thought, in those brief moments, the melodies of *Swan Lake* brought back her inner choreography. How was this possible?

By some act of neurological grace, the cognitive ability to enjoy music remains remarkably intact, even in people who are losing touch with their environment.

Music and memory are entwined in ways that scientists have only started to unfurl. A 2021 pilot study found that regular listening to music that holds personal meaning—like the song someone danced to at their wedding—may stimulate vital brain connections in people with mild cognitive impairment or early Alzheimer's. For three weeks, fourteen adults with cognitive decline (six musicians and eight non-musicians) listened to an hour a day of personally meaningful music. After this three-week period, participants had brain imaging while listening to music they had heard for the first time just an hour before. The scans showed activity mainly in the auditory cortex. But when the same adults listened to long-familiar tunes, they also showed significant activity deep in the prefrontal cortex, a sign of executive cognition perking up. Strikingly, they were more able to recall words such as "face," "velvet," "church," "daisy," and "red" in the right order in a standard memory test.

"Typically, it's very difficult to show positive brain changes in Alzheimer's patients," said Michael Thaut, a senior author of the study and director of the University of Toronto's Music and Health Science Research Collaboratory. Thaut, who holds a Canada Research Chair in music, neuroscience, and health, emphasized that follow-up studies are needed. Nevertheless, results from this pilot study indicate possible improvements in the "integrity of the brain," with regions of the brain "talking to each other more again."

We tend to think of memory as the ability to recall someone's name or remember what we had for lunch. But that's just short-term memory, often the first to go. We also have long-term semantic memory (general knowledge), motor memory (how we remember to ride a bike), and emotional memory, among other kinds, stored in different yet interconnected brain regions. Rhythm and song stimulate many types of memory at once, said Stuart MacDonald, a psychologist at the University of Victoria who studies aging. Music is "incredibly resistant to forgetting."

MacDonald took a closer look at music and memory in an intergenerational choir called Voices in Motion. Based in Victoria, British Columbia, the group teamed up people with dementia, their caregivers, and singers from a local high school.

To his surprise, choir members with dementia were "absolutely able" to learn new songs and recall them week to week, likely relying on types of memory less damaged by dementia.

After fourteen weeks of rehearsals in the pilot choir, singers with dementia showed a slight improvement in their verbal recall. Although music cannot halt the progression of dementia, MacDonald said, he theorized that singing in a choir might have lowered people's depression and stress, "freeing up some of their available [brain] resources so they actually show a bump in terms of their cognitive function."

Lorelle Seal signed up for the group with her eighty-eight-year-old mother, Gracia, who had sung in a choir as a child and played violin. Diagnosed with dementia a year earlier, Gracia could no longer remember people she saw regularly or cook for herself, "not even macaroni and cheese." Yet she had no trouble learning songs she'd never heard before, including the Josh Groban hit "You Raise Me Up."

The elderly singers and high school students enjoyed chatting during coffee breaks, said MacDonald, adding that despite cognitive decline, people with dementia still "laugh, they sing with amazing fervor, they share stories."

This describes my parents to a tee. Russell, at seventy-nine, can't remember what he did half an hour ago but up until my parents' recent move, he never forgot to put on his dancing shoes on Friday nights and head to the pub around the corner in the village my parents called home (population: 200; 250 on pub nights). Although neighbors noticed their memory loss and confusion, pub regulars kept telling me how much zest my parents brought to the party. Starting with a round of beers with friends, they'd sashay to a live band playing rousing renditions of "Mustang Sally."

At some point in dementia, pub nights come to an end. Still, the beat goes on. To connect with family members in the later stages, the Mayo Clinic recommends playing tunes from happier times and encouraging elders to sing, clap, hum, or tap their feet. The change in mood can be dramatic.

Music and Memory, the dementia program featured in the 2014 documentary *Alive Inside*, became the focus of a massive 2020 study conducted in 265 California nursing homes. Staff members created a playlist for each resident tailored to the person's musical tastes. After three months, the residents' odds of needing anti-anxiety medications declined by 17 percent, and aggressive behaviors decreased by 20 percent. All from recorded music alone.

Most of us never picture ourselves in a nursing home with dementia. I certainly don't. But if my time comes, please give me music.

As for my parents, they're slowly adjusting to life in a care home. Russell enjoys the dart games and movie nights, but Mom misses her walks by the ocean and the salad greens she used to harvest in the village garden. I call them regularly to see how they're doing. The other day, one of the staff members sent me a video of my parents listening to sock-hop music in the lounge. They were dancing.

THROUGH ALL THE CHANGES they've faced, my parents have remained good-natured, still cracking jokes. Yet with every memory they lost, they drifted further away from me. Another link to my family disappeared.

Of all my siblings, only my older sister remembers our home in Quebec, the rigors of the music conservatory, and our little cat Minou. My half brother was eleven when I moved away for university and my half sister was barely two. Once my parents are gone, who other than my big sister will know me from the start?

I understood why older people get obsessed with genealogy, searching for connections with relatives long dead. As Alex Haley wrote a year after publishing his epic novel *Roots*, "In all of us, there is a hunger, marrow deep, to know our heritage—to know who we are and where we came from. Without this enriching knowledge, there is a hollow yearning."

For much of my life, I had blamed this hollow yearning on my unfinished business with music. But some of this emptiness, I realized, might have come from living far away from family and moving around too much. In bouncing between music traditions, I had reinforced the pattern. Maybe with music as well as family, I could do a better job of nurturing what I had.

My husband, too, was grappling with midlife. One night he proposed a family "gap year" abroad, hoping to recapture the thrill of travel he'd discovered in his youth. My mind went straight to the consequences for our savings and jobs. *How could we pull it off?* Scott had it all figured out. After months of cajoling, I agreed to the plan provided we make music the focus of some of our travels and get better acquainted with my Eastern European side. Scott left his business in the hands of a partner, while I talked my way into an unpaid leave from work.

The year our boy turned ten, we set off.

— 10 —

Fumbling Towards Ecstasy

IN THE STORYBOOK CITY of Krakow, Poland, my family and I spent Christmas with relatives of the Ukrainian father I had never known. His sister, my Aunt Ksenia, had laid eyes on me only twice before. The first time was a two-hour visit when I was seven. (The Soviet Union granted her permission to travel to Canada, but only if her family stayed behind.) Our second visit lasted three days. This time, I was nearly fifty but felt like a child as she wrapped me in her arms, squeezed my cheeks, and said, "Adrianka, Adrianka" again and again.

Her apartment in Krakow's historic district had herringbone floors, windows draped in mustard-yellow brocade, and a decorative heater that had once warmed a chamber in the fourteenth-century Wawel Castle nearby. The dining room was filled with close relatives I hardly knew: my elderly aunt and uncle, first and second cousins, and a gaggle of bright-eyed children, the youngest being Yurek, a plump two-year-old named in honor of my late father.

Aunt Ksenia's husband, Krystoff, handed me a wafer the size of a playing card that was embossed with a picture of Jesus. Roman, my towering first cousin, instructed me to break off tiny pieces and trade them with each family member while exchanging blessings for the coming year. Many relied on Roman to interpret their good wishes for me. The wafers, called oplatek, tasted as insipid as altar bread, but I savored this sweetness.

When my uncle passed sparklers around the room, everyone began to sing. Polish Christmas carols, I learned later, are an artform of national pride: the oldest can be traced to 1424. During centuries of invasion by the Prussian, Austrian, and Russian empires, Polish composers kept writing new songs to boost morale, penning lyrics about Polish comfort foods and the travails of winter. To this day, singing carols around the table on Christmas Eve is a tradition shared by nearly 80 percent of Polish people, whether they can sing in tune or not.

My family's festive songs had a lilting, folksy quality, and I was charmed by their voices and the children's giggly delight. I had spent my adult life exploring the music of lands unfamiliar to me, without ever knowing about the musical traditions of my own heritage. My heart was full, even before we sat down to eat. And eat. And eat.

Aunt Ksenia, in her late seventies, had spent weeks cooking for Wigilia, the Christmas Eve feast of twelve courses, one for each of Christ's disciples. We tucked into steaming bowls of borscht with mushroom-stuffed dumplings, followed by four different kinds of pierogies, some with fresh cabbage, others with sauerkraut. Then carp, two ways: fried or cold in a squelch of jelly. (Many Eastern Europeans grow up haunted by memories of carp swimming in the bathtub in the days before Christmas.)

By the eighth course, everyone admitted defeat on the supper front and bundled up for midnight mass. Uncle Krystoff ushered us on a snowy walk to his favorite of Krakow's more than a hundred Roman Catholic churches. Each one, he told me, would be standing-room only that night. The Church of St. Francis, inaugurated in 1269, had brilliant blue vaulted ceilings gilded with stars, and walls adorned with paintings of blooming roses and violets. Incense mingled with the tangy scent of Christmas trees propped near the altar and entryways. The bishop wore an embroidered white habit and a folded miter hat, in a style I'd describe as "Pope lite." As he droned on in Polish, I worried the unintelligible sermon might go on all night.

Before long, the organ rumbled and the congregation burst into song. Voices, through sheer numbers, filled the cavernous hollows more gloriously than I had ever heard in a church before—somber, unified, magnificent. Over centuries, songs like this must have won over countless souls. Even though these echoing voices would never make me a believer, they moved me.

Celebrating Christmas with my Polish-Ukrainian family helped me see for myself that music really is the language of culture, and ultimately, kin. The reunion drove home something else, too: My feelings of being adrift weren't just tied to music but also entangled in deeper questions of meaning, family, purpose, and belonging. At the core, my issues were spiritual.

SPIRITUAL NEEDS, LIKE MUSIC, may be hardwired in us. Roughly a hundred thousand years ago, human remains were buried in Qafzeh Cave, Israel, with tools marked with red ocher. The first evidence, perhaps, of ritual behavior.

Spiritual beliefs may arise from the brain's oldest and innermost region. Using a technique called "lesion network mapping," researchers at Harvard University charted the religious and spiritual beliefs of brain-injured surgical patients and Vietnam veterans. They traced these beliefs to the periaqueductal gray, a circuit in the brain stem that is sensitive to pain, fear, compassion, and love—and also highly responsive to music. Spiritual behaviors may be "deeply woven" into our neurobiology, said the study's lead author, Michael Ferguson, and "centered in one of the most evolutionarily preserved structures in the brain."

Our openness to spiritual teachings, then, might stem from primal yearnings for safety, connection, and love. These needs haven't changed. People who feel a sense of "oneness"—the feeling that "everything is connected"—report higher happiness and life satisfaction than those who do not. In a 2019 study from the University of Mannheim, Germany, this link persisted whether a person was Christian or Muslim, atheist or agnostic.

Findings like these made me wonder if I could cultivate a state of grace without meditation, religious philosophies, or new-age beliefs. For much of my life, I'd been searching for an organizing principle to live by, a way to dissolve myself into something infinite, rapturous. Only recently had it occurred to me that I might enter this state through music.

For those of us raised without religion, it is easy to overlook the central role of music in the world's spiritual practices. But imagine a Baptist church without a choir. An Indigenous Sun Dance without drumming. A synagogue without a hazzan (cantor) chanting Hebrew prayers. Buddhism might seem the exception, with its silent meditations, yet many Buddhist traditions use rhythmic chanting to prepare the mind for contemplation. In Tibetan Buddhism, the journey to enlightenment can be cacophonous, filled with growling overtone singing, ritual bells, cymbals, oboes, and nine-foot-long trumpets called dungchen, likened to the blasts of elephants. Worldly attachments, begone!

Music primes us for spiritual teachings, reaching us in ways that sacred texts alone do not. Brain scanning at McGill University has shown that melody is processed in the right auditory cortex, while speech is processed in the left. Words put to melody light up the auditory regions in both hemispheres and in our emotion centers, turbocharging the message. No wonder hymns are easier to remember than scripture, and songs carry special meaning in cultures worldwide.

In my mind, an underlying question remained: Could music alone offer a direct path to oneness?

My search for answers led me to an experiment with hallucinogens, a journey to the birthplace of ritual music, and conversations with scientists mapping the frontiers of consciousness and the human soul.

SHORTLY BEFORE MY FAMILY set off on our gap year, I went on a different kind of trip.

Ayahuasca, the psychotropic "tea," had intrigued me ever since a pilgrimage in my mid-thirties to Cuzco, Peru, the Incan capital that had become a launchpad for Machu Picchu tourists. As I hiked the Inca Trail, a pair of awestruck travelers told me the brew had conjured visions and insights unlike any they'd had before. It also made them vomit their guts out. At the time, despite my existential angst, I didn't feel safe entrusting my insides to a roadside healer in Peru.

Years later, in Canada, my curiosity grew when a friend mentioned the medicine songs known as ikaros that accompany an ayahuasca ritual. Ikaros aren't just a soundtrack for a drug trip, she said, but an essential ingredient in what she described as a transformational experience. I had to know what an ikaro sounded like while under the influence.

My friend put me in touch with a Canadian curandero who had trained extensively with a master healer in South America. After a year-long wait, I joined about forty other adults in a wood cabin overlooking the northern rainforest.

Ayahuasca is made by boiling the leaves of a tropical shrub with the bark of a rainforest vine that grows as thick as a forearm. That night, I swallowed this vine of souls from a shot glass filled with dark brown sludge. It tasted acrid, yet cloyingly sweet, like molasses mixed with motor oil. I could barely get it down. At first, I felt nothing. As I lay on my Therm-a-Rest, a hand's width from neighboring mattresses, I heard someone sobbing at the other end of the room, followed by the sound of puke spilling into a bucket. Ugh. It was going to be a long night.

Then I heard the ikaros chanted by the healer and his attendants—guttural, otherworldly, unlike any sound I'd heard before. The mewling, whistling tones seemed to apparate out of the ceiling and slither into the floor, the auditory equivalent of Salvador Dalí's melting clocks. I felt a whirring in my leg, as if a tiny motor were lodged inside. The eerie songs had penetrated my skin, setting more tiny motors vibrating at points throughout my body. Numbed tissues sparked to life.

The healer's mouth appeared to me as a gaping maw, with intricately patterned snakes slowly wriggling between his teeth. I watched in wonder, neither frightened nor disgusted, as the magical creatures slid towards me. Then the vision faded, leaving me vibrating into the wee hours in a figment of space and sound.

Ikaros, I read later, are improvised, passed down from one healer to another, or learned during a period of isolation from human contact. The healer ingests a specific plant to commune with its essence and learn its song—a way to communicate with a being that does not understand spoken language. Bernd Brabec de Mori, a musicologist at the

University of Innsbruck in Austria, worked among the Shipibo-Konibo people in the Peruvian lowlands for more than half a decade. Healers combine songs depending on their intentions, he wrote, which might include summoning plant or animal beings for guidance or scaring away dark energies.

Vibrant tapestries embroidered by the Shipibo people have been likened to musical scores, or "singable designs" that illustrate the patterns of medicine songs. Brabec de Mori says this notion is pure myth, invented in the '80s by a German anthropologist. I don't doubt his scholarship. Yet as I scrolled through Shipibo textiles online, one of them gave me a start: it was the spitting image of the geometric background I'd seen in my ayahuasca vision of snakes, interlocking and infinite, like an M.C. Escher drawing. How could this be?

I thought of cymatics, the study of visual patterns made by sound. If you put a pinch of salt on a flat surface attached to a loudspeaker, sound vibrations will scatter the salt grains into shapes that look like snowflakes or kaleidoscope designs. Does ayahuasca allow the brain to "see" the vibrations of Shipibo songs? Who knows.

Ayahuasca is a powerful drug, shown in a study at Imperial College London to induce a dream-like state while stimulating "chaotic patterns" of brain activity. In many Amazon communities, before the explosion of ayahuasca tourism, only healers drank the brew because they alone had the training to grapple with demonic apparitions or interpret lucid visions. Experienced users warned me that ayahuasca may trigger nightmares or fantastical dreams that recur for months. I had gotten off easy in my first experiment (and likely my last, as a metaphysical path lined with terrors isn't for me).

Two months later, though, I dreamed I was dancing while experimenting with different sounds on an Amazonian flute.

A stranger walked up and asked me to play for him, but I choked. Then I tried again, this time making sounds more ethereal than before. Little sparks and puffs of smoke shot out of the flute holes, alarming me, but I was too mesmerized to stop. In a flash, the outline of a giant turtle materialized above me like a zodiac constellation. Lines of searing white energy radiated from the flute, tracing their way around the turtle with blinding speed, frightening me. Knowing something momentous would happen once the pattern was complete, I dropped the flute, shutting everything down.

I awoke from the dream annoyed with myself. *Why had I dropped the magic flute?* Just like when I was a child, I'd played beautifully, only to dismiss it. My thoughts turned to the giant turtle. What if I had absorbed its luminous power instead of blocking it?

Weeks later, I stumbled upon archaeological research confirming what Indigenous peoples have always known: turtle shells uncovered in sites across North America were more than just food scraps. As musical rattles, these turtle shells imbued dances and ceremonies with the rhythms of creation—and the symbol of how the world was formed on a turtle's back.

Then I learned that in Greek mythology, the first musical instrument was created when the god Hermes gutted a tortoise to create the soundbox of a harp-like instrument called the chelys (turtle).

Never before had my mind conjured ancient turtle myths, South American snake spirits, or geometric Amazonian textiles. Somehow I had tapped into the collective unconscious, in visions that seemed driven by ayahuasca and music. But in the volatile chemistry between the human brain and psychotropic plants, how do otherworldly songs fit in?

INDIGENOUS PEOPLES typically combine mind-altering plants with music. Both have ancient roots and strong ties to spirituality.

In the Southwest of America, the Navajo eat the bitter tops of cactus while singing peyote songs to the beat of a drum made of deerskin stretched over a partially filled kettle. In the mountains of Oaxaca, Mexico, the Mazatec Indians hunt for mushrooms with droopy caps that taste of cucumber. The fungus is eaten fresh, for divine communication, but never without a curandero singing sacred songs by candlelight.

Across the Atlantic, in central Africa, the Mitsogho people of Gabon consume the psychoactive root bark of the Iboga shrub, in ceremonies that offer near-death experiences and spiritual rebirth. Throughout the ceremony, musicians play rattles, a harp, a mouth bow, and ankle bells. Music is the lifeline that tethers the initiate to the world of the living, even as their spirit travels to the hereafter. German scholars said the jingling sounds helped them cling to their mental well-being during the harrowing ritual "death." At the same time, they wrote, music reactivated the out-of-body visions, which tended to ebb during moments of silence.

This reminded me of my ayahuasca trip. When the chanting stopped, the visions stopped. Everything stopped.

Mendel Kaelen, a Dutch neuroscientist, has been studying the mental and spiritual benefits of music combined with psychedelics. Now in his mid-thirties, he became fascinated by altered states after reading Robert Monroe's 1971 book, *Journeys Out of the Body.*

At Imperial College London, Kaelen and colleagues discovered that a single session of psilocybin ("magic mushrooms") lifted depression for as long as three months in seven out of twelve people who hadn't improved with antidepressants or talk therapy. Kaelen observed something else as well: music, routinely played while patients were tripping, seemed to amplify the therapeutic effects. (He wasn't the first to notice. In 1965, Abram Hoffer, a Canadian physician studying LSD as a treatment for alcoholism, described music as "very useful" in bringing out the psychedelic reaction.)

In a series of studies, Kaelen confirmed that the more "in tune" patients felt with the music while on psilocybin, the more likely they were to have a mystical experience. More importantly, the quality of the musical experience strongly predicted improvements in depression, while the general intensity of the drug trip did not. Music acts as a "hidden therapist," said Kaelen, increasing the chances of psychological and spiritual transformation.

Melodies and rhythms shape the drug experience by evoking strong emotions and mental imagery. A rising melody, for example, might trigger the perception of being "lifted up." Music "can almost provide a sense of a narrative that you travel through." At the same time, music helps keep the brain on drugs from spinning in all directions, potentially missing out on insight and psychological growth.

A 2018 paper in the *International Review of Psychiatry* concluded that together, music and psychedelics interact to produce "profound alterations in emotion, mental imagery, and personal meaning." (A word of caution, though: participants in these studies were under medical supervision; tripping to music at a party or alone might not offer the same results.)

Kaelen now believes music can act as a psychedelic on its own: "Music reveals the soul." He became so convinced that he quit his university job to focus on Wavepaths, a start-up that aims to offer insightful experiences with or without psychedelic drugs.

Wavepaths uses artificial intelligence to tailor music to the listener's mood, based on vital signs such as breathing. But unlike most music apps, Wavepaths doesn't supply premixed songs. Instead, generative algorithms turn ambient tracks by artists such as Brian Eno into ever-evolving compositions that never sound the same twice. The goal is to avoid unwelcome associations with familiar music, increasing the chances of resonance for the listener.

Kaelen's description of Wavepaths reminded me of the meditative "sound bathing" I explored in chapter 7, except instead of touting "cosmic vibrations," Wavepaths draws people to music with technological bells and whistles. I'll be curious to follow his work to see if they really add much.

In Indigenous rituals worldwide, healers intuitively adjust their rhythms and songs to meet the participants' moment-to-moment needs. They don't need software or AI to transform the human psyche.

WAREHOUSES PULSATE WITH PEOPLE dancing for ten, twelve, or sixteen hours straight, sweat streaming from their bodies as strobe lights flash in time to the beat. The music is loud and fast-paced, mixed by a deejay to seamlessly morph from track to track, creating patterns that one writer described as a "fractal mosaic of glow-pulses and flicker-riffs." With or without party pills like Ecstasy, dancers abandon themselves in a kind of "ego-melting mass communion." Drugs may enhance the experience, but they don't define it. For ravers, the main draw is the music and the shared values of PLUR—Peace, Love, Unity, Respect.

I've never been to a rave (even with earplugs, heart-pounding electronica is too intense for me). But I can understand the appeal of losing oneself in oceanic waves of sound. Robin Sylvan, an American religious studies scholar, believes events like raves fulfill deep-rooted needs: "Music is something that induces these very powerful experiences, life-changing experiences," he said, "and they're clearly religious and spiritual in nature."

Sylvan immersed himself in communities of ravers, Grateful Dead fans, hip-hoppers, and metalheads in scholarly research that led to his book *Traces of the Spirit: The Religious Dimensions of Popular Music*. Like organized religions, musical subcultures offer a sense of belonging and identity. These rituals, though secular, offer "an encounter with the numinous" that is central to all religions. Raves focus on a very strong hypnotic rhythm. Even if you just dance without doing drugs, said Sylvan, "you're going to enter into an altered state of consciousness."

By "altered state," he means "trance," a slippery word. It makes me think of clairvoyants, Ouija boards, and crystal healers channeling past lives. Our science-revering society lacks social norms for altered states that don't involve drugs or alcohol. Psychiatry has taught us to think of our minds as gatekeepers with a mission to tame our amoral id. If trances are real (and for many people, that's a big "if"), we don't see an upside to mental states beyond our conscious control.

But that's just our hypervigilant prefrontal cortex talking. The truth is, most of us slip into a light trance several times a day, from daydreaming to absentmindedness to the suspension of reality that allows moviegoers to forget they're watching actors on a screen. (Lance Rucker, a retired dentistry professor, told me patients often slip into a light trance during dental procedures. That's one way to get through it.)

Only at the far end of the spectrum of mental perceptions do we find extreme altered states, from hypnotic trance (detectable in brain scans, believe it or not) to a complete break from reality.

In trance rituals worldwide, the driving force is drumming. But how does music put us in a trance? Music-induced trance is tough to study. After all, lab environments are worlds apart from the crackling fires, vibrating drum skins, and swaying dancers that often accompany trance rituals. Nevertheless, in a 2015 study, a cognitive neuroscientist at Harvard University found a way.

Michael Hove recruited adults trained to enter a trance at will using techniques taught by the late anthropologist Michael Harner. Participants lay in a cylindrical fMRI scanner while listening to repetitive hand drumming. In a series of eight-minute brain scans, they were instructed when to go into a trance or not. (For the non-trance runs, the drum track was given irregular timing to reduce the chances of an unintended trance.)

All participants said they entered an altered state in the trance sessions but not the non-trance ones. Their noggins confirmed it. During the trance runs, their brains lit up in networks associated with an internally oriented cognitive state, and darkened in areas involved in sensory processing. The study described repetitive drumming as "a powerful method to alter consciousness" used since ancient times, and concluded that trance-inducing techniques such as drumming "likely have a common biological basis."

Some scholars regard ritual trance as the oldest form of spirituality, predating all known religions. If they are right, events like raves may not mark a departure from age-old spiritual practices but rather, a return.

Despite several years of drumming—and many sublime moments in the studio—I still didn't know how to enter a state of oneness at will. I wondered what I might learn from a musical culture whose spiritual life revolves around rhythm and song.

IN THE GARDEN of Small World Backpackers Lodge in Harare, Zimbabwe, I spoke with a traditional healer and musician,

Caution Shonhai. A tall man in his early fifties, he chuckled when I asked if his personality matched his name. "Yes, very much so," he replied with a wide smile. We sat under the shade of banana trees bursting with purple pods the size of bovine hearts. Then he pulled out his mbira—the ancestral instrument of the Shona people—and played a song for me.

The mbira is pronounced "m-bee-rah" with a rolled "r," and has two dozen metal tines fastened to a thick slab of hardwood. An mbira, depending on the player, can growl, bellow, quaver, weep, or tinkle light as rain. As Shonhai sang in the Shona language, the gentle melody flowed like water over pebbles in a stream. I could have listened for hours, just drifting away.

Shonhai often plays this song, "Chigwaya," for patients in the hospital. Mbira music "goes to your brain," he explained, "and then from the brain it will help the blood to flow calmly, so that the body will be also." The lyrics roughly translate as "the bream fish always enjoys being in its pool." Like a fish in water, he said, this song tells us to take it as it is: "You have to be satisfied with what will be taking place within your body, or within your life."

During my visit in 2019, I got a glimpse of the challenges Zimbabweans faced, from mile-long queues for gas to bread prices tripling in a month. The Republic of Zimbabwe, formed in 1980, is still reeling from the impacts of nearly a century of colonial rule. Shonhai told me he plays mbira every day, reaching for the instrument whenever he feels a weight on his shoulders. He'll play for an hour or so, often in the middle of the night, "and then I feel like, oh, I am now free."

Every mbira player I spoke with said the music connects them with the spirits of their ancestors. (I interviewed seven professional mbira players and had informal conversations with a dozen others; see notes.) In Zimbabwe, I saw people playing mbira at the bus stop and in city parks. Small enough to fit in a purse or backpack, this iPad-sized instrument allows them to communicate with the source of all creation in any given moment.

The mbira dates from the Later Iron Age, about a thousand years ago, and possibly much earlier with tines of reed. In the twentieth century, British colonists called it a "thumb piano" and marketed it outside Africa. One ethnomusicologist took an instrument from the mbira family of lamellaphones and gave it Western tuning—along with his own name: the "Hugh Tracey Kalimba." (Today we call it appropriation.)

In their original tunings, the metal tongues of mbiras whisper ephemeral melodies. Unexpected overtones buzz from the bottle caps and metal beads wired to the mbira and the giant gourd used to amplify the sounds. When multiple players get together, interlocking melodies combine in a "helix-like weaving," said Martin Scherzinger, a mathematician at New York University who has rigorously analyzed mbira songs. Off-kilter patterns, starting one beat apart, combine in uncanny textures and "melodies that appear as phantoms."

In 2020, UNESCO inscribed the crafting and playing of mbiras in its list of the Intangible Cultural Heritage of Humanity.

I'd learned a few mbira songs after taking workshops in Canada. Did an outsider like me have any right to play this sacred instrument? When I asked Shonhai, he looked at me as if the answer were obvious. While he emphasized that he only spoke for his own clan, he said, "my instrument is your instrument. That is my belief. Because we came from Tanganyika," he explained, using the historical name for Tanzania. "Both of us."

His knowledge of humanity's shared origins shouldn't have surprised me. From what I have gathered through conversations with Shona people, ancestral connections are at the heart of the cultural practices, philosophy, and spiritual beliefs known as Chivanhu.

Long after the bodies of our parents and grandparents have passed away, said Shonhai, "their spirits are within us." I asked if he was talking about genetics: one could say that at a DNA level, our ancestors live in us. Yes, Shonhai assured me, he knew about genetics. But for him, ancestral spirits aren't just a metaphor for something else. His body is regularly claimed by ancestral spirits, he said, including that of his great-grandfather, Nzuwa, a master mbira player. Although he has no memory of being possessed during these sessions, ancestral spirits give healing advice to his community.

I nodded, not sure what to say. I'd never seen anything like what he described.

Mbira music, he continued, entices ancestors back to the world of the living, acting as a "cell phone to the spirit world." He smiled when I failed to hide my disbelief. "Most people, they believe that if the body dies, even the spirit vanishes," but "in our culture, because we are very lucky, we can even call our ancestor, or ancestral spirits, and they can possess... people. And we can talk direct." His eyes shone as he spoke. I could see that for him, this connection was as real as the ripe bananas were sweet.

Shonhai added that wherever he has traveled to teach music and perform, from California to Illinois, people have asked how the mbira might help them connect to their own spirituality. "I always tell them to sing in their own language, so that it will be easy for their ancestral spirits to understand."

I could see the value in honoring earlier generations who helped shape the current iterations of "you" and "me" (for research in this vein, see notes). But in my mind, connecting with family history was miles apart from a belief in out-of-body communion with ancestors.

When I planned a visit to Zimbabwe, I thought I could learn about the mbira and Chivanhu traditions without encountering spirit possession. I was wrong.

MY JOURNEY TO ZIMBABWE began at a music festival on Vancouver Island, British Columbia, where an mbira teacher and scholar, Moyo Rainos Mutamba, told me about the intentional community he and his family had established several hours' drive south of Harare. They'd named their village after the African philosophy of Ubuntu, which loosely translates as "I am, because we are." Mutamba's family had reclaimed ancestral spirituality, along with the mbira, denounced to this day by some Christian groups as an instrument of the devil. As we chatted, I hatched a plan to visit Ubuntu. I wanted to meet people who thought of music as a force of nature, essential to life.

On the long, bumpy road from Harare, my husband, son, and I passed crumbling townships and a giant billboard warning of AIDS. During a stop at the supermarket, I watched people buy groceries with a stack of bills as thick as a brick. I could imagine the stress of runaway inflation and the fear of being halted at a checkpoint by armed police. As soon as we rolled into the village, though, I began to unwind. Round huts stood in a cluster, wearing cone-shaped hats of thatch. Dogs lazed in doorways, ignoring the flies on their coats. Chickens pecked at the ground, followed closely by tiny chicks, white and black. Beyond the earthen huts, fields of maize, onion, potato, and covo, a type of kale, spread as far as I could see. (I would learn about these crops while harvesting them for supper.)

Villagers ushered us into the kitchen hut and showed us our seats on the soot-blackened floor around the fire. Then each picked up an mbira and began to sing. As they plucked the handcrafted metal keys, the gentle murmurs of one mbira mingled with the bass notes of the next. Bell tones from another melody chimed in, adding to the echoing layers. The mbira players smiled and shook their bodies in time with the rolling triplets as their fingers and thumbs flew over the keys. Some of the women trilled their tongues from side to side in shrill ululations, as if to say "Ramp it up!" Others sang syllables like "ya-ya-ay" and "we-he-wah." It was uplifting, hypnotic. Long after my family went to bed, tired from hours of travel, the trilling melodies pulsed through the night, shooting out the door of the kitchen hut and up into the stars.

Over the next few days, I joined the women weeding the fields, cooking meals, sweeping the grounds, washing dishes in muddy water. Many of them had a cell phone, charged with a solar-powered car battery, and online accounts on Facebook and WhatsApp. Yet they spent most of their free time dancing and singing, belting out songs generations had sung before. Mbira music made them happy, they said, and helped them dream at night.

On the third day of our stay, a man in the village beckoned me for a session with the visiting healer, no payment asked. I saw no reason to refuse. After all, I'd traveled far to immerse myself in experiences unfamiliar to me, from spotting dried caterpillars in plastic packets at the supermarket to hearing music said to bring back the dead. I followed the villager into a boxy concrete shack.

Inside the dark room, the air was peppery with snuff. I stared at the healer hunched on a bamboo mat in the corner. The day before, the same man had introduced himself as Jonathan Goredema, a former engineer. Broad-shouldered and bright-eyed, he'd told me about the plants he harvests from the woods to treat fever or stomachache. Today, though, the man before me looked contorted, as if in pain. His eyes were partially closed and his face—though I couldn't wrap my head around this—seemed to have aged decades.

Covering his legs was a length of red-and-black fabric with a geometric pattern of squares and dots, which I recognized as the retso design, said to honor the shave rembira (spirit for playing mbira). An hour earlier, I'd heard tinkling melodies flowing through the door of this shelter. The ancestors' favorite music. Had they answered the call?

Jonathan had left the building, I was told. The "old man" possessing him sat between a young family member and a slender assistant named Alois Mutinhiri. Unlike Jonathan, who spoke perfect English, the ancestor inhabiting him communicated only in the Zezuru dialect. Mutinhiri had to interpret everything the spirit said. He asked if I had a question for him.

At first I couldn't think of anything. Moments later, words tumbled out of my mouth. "My body is tight, my mind is tight. Constricted," I said. "I don't know what to do about it. I've tried so many things." My stomach clenched.

Wordlessly, the spirit-man reached for a small wooden bowl and snorted a pinch of snuff. Then he cleared his nose and mumbled something in Zezuru, breathing heavily. "This tightness is not your pain," Mutinhiri interpreted. "Your ancestors have suffered. Long, long time. Way back."

The family history I'd learned on my trip to Poland the month before was fresh in my mind. Welling up, I talked about my father, his death from cancer at twenty-nine, and the death of his thirty-three-year-old mother—my grandmother—when he was two. Soon after his mother's death, my father was abandoned by his father, who died in midlife. So many losses, against the backdrop of atrocities committed at Auschwitz, less than an hour's drive from my aunt's apartment in Krakow. In that moment, the weight of all these deaths overwhelmed me, practically crushing me into the concrete.

The men watched me in silence. Then the spirit spoke. My ancestors were in conflict in my body, he declared, but "they must work together." He told me I needed to separate myself from their suffering.

"How?" I asked, looking down at the floor.

The spirit offered to talk to my ancestors. If you go back far enough, he assured me, his ancestors and mine were the same. I pondered this. Here in East Africa, the cradle of humankind, the idea of shared ancestral spirits made perfect sense. "Thank you," I said, comforted. But there was more.

"You have so much fear. You are afraid of everything."

"Yes, I know." I gazed at the grimy window above the men's heads, not knowing where else to look. A pain had been throbbing in my forehead since the celebration in the village the day before. Each time I'd tried to join in the dancing and drumming, my every move felt stilted, stiff. I was furious with myself. I knew how to dance, how to drum. *What was wrong with me?* The rhythms in the village were so strong that nothing I did could wreck the moment for others. *What was I afraid of?*

In the dark of the shack, I broke into sobs. The ancestor paused and then said something I didn't expect. "Very soon," he said, slowly, "your talent will be known. You will succeed."

Succeed in what? I had given up being talented, in music or anything else. My only goal at the time was to finish writing a book—this one—and I couldn't even decide what story to tell.

"I don't care about success anymore," I replied.

The spirit wasn't buying it. "Your work will be known," he said. "This will happen soon. You must have courage." Mutinhiri grinned as he translated, shaking his head. "No more fear."

I thanked the spirit, unsettled by this pep talk from the beyond. That afternoon I sorted dried beans with one of the women, laughing when she kept tossing the worm-eaten ones back into the pot. "Is good!" she said. My body felt lighter than it had in days. Catharsis, I figured, from my breakdown in the shack.

I chatted with the women long into the night, enjoying the tropical air punctuated by the squawk of turkeys. Gazing into the starry sky, I thought of the village in Chiapas where I'd stayed when I was small, playing with my mother's clay flutes and nestling into her body whenever I could. That village, I imagined, must have felt a lot like this.

MY ENCOUNTER with a healer possessed by an elderly ancestor had no easy explanation. He didn't charge a fee or ask for publicity, and like my husband, he had earned a livelihood in the exacting discipline of engineering. I couldn't see a reason for this guileless man to fake spirit possession.

The idea of someone's mind and body being taken over—a complete loss of self—didn't square with my version of reality. Yet belief in spirit possession is more common than I knew, in a wide range of groups, including Catholic and Jewish communities. A mid-twentieth-century survey of 488 distinct societies found that 74 percent had practices involving trance or spirit possession. To this day, many believe someone can be inhabited by a spirit, whether Satan, the Holy Ghost, or the malevolent jinns of Islamic mythology.

Spirit possession, a form of trance, is most often driven by music.

In a Haitian drumming ceremony, the anthropologist Wade Davis witnessed a petite woman gain superhuman strength. The initiate, once possessed, could lift heavy men off the ground and "swing them about as if they were children." As the drumming reached fever pitch, she took a red-hot coal "the size of a small apple" and balanced it between her lips. But the first time I read this account, I could hardly believe it. Davis acknowledges that for outsiders, "there is something profoundly disturbing about spirit possession."

Pentecostal Christians, among other communities, describe it as a blessing.

Pentecostals sing and dance ecstatically to Gospel songs until the spirit of God enters their bodies, uttering divine languages that mere mortals cannot understand. Andrew Newberg, a neuroscientist at Thomas Jefferson University, has scanned the brains of Pentecostals who practice glossolalia, "speaking in tongues." (For Newberg's method, see notes.) During glossolalia, Newberg saw a drop in the frontal lobe, a brain area that helps us feel in charge of what we are doing. This finding, he said, mirrored the Pentecostals' experience of their conscious selves being "taken over" by the divine spirit.

In other studies, Newberg has observed a distinct brain pattern in devotional practices that involve "self-surrender," including rhythmic movements, chanting, and prayers performed by Sufis to achieve a "mystical union with the divine." Brain activity dropped in the frontal lobe and the parietal lobe, a region near the back of the head that helps us orient ourselves in three-dimensional space. Newberg speculates that these decreases might explain altered states that bring feelings of vastness and a blurred sense of self.

Skeptics have argued that sensory perceptions involve multiple brain areas at once, and cannot be pegged to a specific region or two. Nevertheless, at the very least, Newberg's work suggests that when a Pentecostal Christian or a Zimbabwean spirit medium says they've been possessed, something measurable has happened in the brain.

Intense experiences of self-transcendence have been linked to altruistic behaviors that may last for months. However, as Newberg points out, "spiritual practices are more likely to offer transcendent experiences in people who fully engage with them." With that in mind, "it is important to choose a practice that aligns with a person's cultural background or their prevailing belief system."

Moyo Rainos Mutamba, the Shona musician who put me in touch with Ubuntu village, told me that when he plays the mbira, he feels the presence of his ancestors all around him. "I am never alone." What a comforting feeling.

As for me, I've come to realize that I lack the belief system that would allow me to embrace spirit possession, Christianity, Tibetan Buddhism, or any other faith. I do believe in science, though. And through music, I'm learning to open myself up to the dreamy state that some scientists call "trance" and others call "flow," enough to enter a space where I can, for a little while, truly let go.

A FEW MONTHS after our trip to Zimbabwe, our family gap year wound down in Provence, in the South of France. It was mid-June. Sun baked the slopes of Sainte-Victoire, the ice-cream-colored mountain that the Impressionist painter Paul Cézanne rendered in oils and watercolors more than eighty times. Below, in the scrubland where I walked, tangy aromas of wild rosemary and thyme wafted in the breeze. The annual cicadas, beckoned by the heat, had crawled out from underground and erupted in song. Or, more accurately, a piercing, relentless, buzzing drone.

I had heard cicadas before, but nothing like the barrage of Provence's more than half a dozen species trilling in chorus.

Male cicadas chirp by vibrating flap-like membranes on their nearly hollow abdomens, amplifying mating calls that female cicadas can hear from as far as a mile away. The cicadas' shrilling membranes, called tymbals, share their name with timbales: high-pitched Cuban drums.

At close range—less than two feet—a cicada's call can hit nearly 107 decibels. Louder than a motorcycle.

I paused on a ledge of pockmarked limestone, in the shade of a contorted pine, and closed my eyes. Even from a safe distance, the insects' buzzing seemed an impenetrable din. But as I listened closer, my ears picked up subtle variations in their chirps. Bands of cicadas sang at different tempos with different accents: *Tzzizzzt! Chhhhrrruzz! Brrrraaeeett!*

The sounds made me think of the buzzing bottle caps attached to Zimbabwean mbiras. Listening to the insect chorus, I began to hear complex divisions of time, in dazzling patterns that reminded me of African polyrhythms. My skin prickled. Then it hit me. The world's loudest insect, the African cicada, must have electrified the soundscapes of primordial humans. Did the very first drummers take inspiration from insects? This cacophony was impossible to ignore.

As I sat and listened, the lineage of human musicking was no longer just a concept, but something I could sense and feel. A kind of oneness.

The ancient Greeks wore golden cicadas in their hair as a symbol of Apollo, god of music and the sun. In the markets of Provence, cicada-themed ceramics and table linens are emblazed with the phrase Lou souleù mi fa canta, Provençal for "The sun makes me sing." Through much of Western history, however, the cicada has been portrayed as a disgraceful layabout. In the famous fable of the ant and the cicada (later, a grasshopper), the cicada sings all summer while the industrious ant toils in the fields. Come winter, the ant feasts while the cicada starves. Moral of the story: work, work, work.

To the ant, music is unimportant, a useless frill. Industrial societies would turn us all into worker ants, snuffing out our cicada nature—the part of us that knows how to dance, drum, and sing. But I believe our health and future as a species may depend on finding a balance between our driven, productive sides and our musical, embodied, sensory selves.

This view is gaining ground. Back in Canada, I would read about musicians in New York State who perform live in fields of cicadas and try to tune in to the surrounding din. Far from crazy, wrote David Rothenberg, a professor of philosophy and music at the New Jersey Institute of Technology, making music with insect hordes reclaims a crucial connection: "Human sounds must fit into and around the callings of nature if we are ever to construct a sure, more promising way to survive on this complex and beautiful planet."

Even if I couldn't see myself jamming with cicadas, I had to agree. Music is an antidote (though not the only one) to a way of being in the world that has us thinking and behaving increasingly like machines.

I drew this theory from Iain McGilchrist's epic book *The Master and His Emissary: The Divided Brain and the Making of the Western World*. McGilchrist, a Scottish psychiatrist, believes the modern brain is dangerously out of balance. Over the past two hundred years, our analytical left hemispheres have reshaped the world in ways that threaten our well-being and our earthly home. "We behave like people who have right-hemisphere damage."

On a walk through the forest, the right lobe is like an Impressionist painter capturing the essence of the experience. The dappling of light on water droplets and glossy spring leaves. Our left lobe, meanwhile, zeroes in on details and categories—species of trees, types of birds, resources to extract. The left lobe excels in accuracy, prediction, organization, and control. But it does not understand beauty, spirituality, nuance, or metaphor.

To be clear, McGilchrist isn't trying to revive the debunked notion of "right-brained" creatives versus "left-brained" Spocks. Both brain lobes contribute to everything we do. "It's not about thinking versus feeling," he explains, "but about two kinds of thinking." We need both kinds.

While the left lobe helps us learn a song in detail, it's the right lobe that transforms patterns of notes and silence, nonsensical on their own, into an expression of what it means to be human and alive.

There on the ledge below Cézanne's beloved mountain, my brain had transformed an aural assault into intricate sounds of intangible value. But I couldn't fully perceive the insects' song until I slowed down enough to listen.

Throughout my Western classical training, I had approached music like a worker ant. Now I'm becoming more of a cicada. I no longer play an instrument to master it but to enjoy the feeling of metal keys buzzing under my fingers and the vibrations of drums penetrating my skin. When I pick up an mbira or stroke a ukulele, I feel a connection to the breadth of human music-making and marrow-deep emotions I sometimes struggle to name.

Music as medicine is an evolving science, but I don't turn to music mainly for the health benefits I've covered in this book. Rhythm and song give me something far more meaningful and sustaining. They offer, in McGilchrist's words, "a sense of beauty and a sense of awe."

AFTER MY FAMILY'S RETURN to Vancouver, I joined a lesson with my new hand-drumming teacher, Alexandra Jai, in her backyard. Birds chirped an evening song. The air had the nip of early fall.

Alexandra started by telling us about a health crisis in her extended family that had left her feeling "so unsettled, so imbalanced." In recent months she had gravitated to rhythms in groupings of five, seven, and nine—asymmetrical numbers common in Central Asia and South Asia, but not in the West. "Odd times call for odd measures."

Her words were both a metaphor and a pun. A measure in music is the number of beats to a bar, the "time signature" of a rhythmic phrase (2/4, 3/4, and so on). In Western music, the most basic measure has two beats, matching how we walk: right, left, right, left. Bipedal style. A waltz has groupings of three: one-two-three, one-two-three. Most often, though, Western rhythms come in multiples of twos or fours (early Christians liked to keep things square). These patterns are so ingrained in our bodies that veering away from them can be profoundly disorienting, as if the floor has been tilted beneath our feet. Everything feels out of whack.

But, Alexandra said, coaxing ourselves out of our rhythmic comfort zone can help us roll with the unexpected in other areas, too. Drumming odd measures becomes a spiritual practice, teaching our minds and bodies how to lean into a new reality. Accept what is.

She taught us a rhythm that would have been simple if not for the fifth beat that threw us all off. Fortunately, the human brain is a master of pattern recognition. After the first awkward rounds, I could feel the rhythm in three beats plus two, and began to groove with the pattern in five. *Cool! Can we play some music now?*

Not so fast. Alexandra taught us a second rhythm in five; shorter but trickier, with bass notes in unexpected places. Half of us played the first long pattern in five while the others fit two cycles of the shorter pattern into the first. Next, she had us switch parts and alternate back and forth. Then she picked up the pace. The beats flew by too fast for anyone to keep track of every beat. The only way to plug into a "five" frame of mind was to let go of conscious control and drum with blissful abandon.

Within half an hour, through rhythm, Alexandra had taught us how quickly we can adapt.

For days afterwards, I felt silly pride in how the members of my body—hands, wrists, shoulders, arms—had "re-membered" themselves to play in fives instead of fours. As I washed dishes or brushed my teeth, the pentagonal rhythms rattled around in my head. I got so pumped that I hummed them to a friend over the phone, raving about the embodied experience of embracing odd measures. Even though she humored me, I knew my words failed to capture the feeling.

Searching for a metaphor, I resorted to math. Most of us can remember a schoolteacher drawing a number line with 1, 2, 3, etc., to the right of the zero, and -1, -2, -3 to the left. Basic stuff. But then we learned about all the "irrational numbers" tucked in between these integers, such as Pi and the square root of two—mysterious numbers with no end. In drumming class, immersed in dizzying rhythms, I can almost feel the tug and pull of those irrational numbers stretching to infinity. While every beat remains precise, the spaces between them seem to morph until I slip into a whorl of circular time. I can feel the Golden Mean. I can feel Pi. It's a trippy, gorgeous, expansive feeling. Ravers call it trance. The Greeks called it ekstasis (ecstasy)—"to stand outside one's self."

SOON AFTER my pentagonal drumming fever, I had a ruby-slipper moment in a dentist's chair. As the drill whirred away at a molar, prepping for a crown, I slipped into a hazy state, perhaps a light trance. My mind meandered from rhythms in fives to the village in Zimbabwe, and all the way back to my childhood home.

Something my mother said when I confronted her about my cello training popped into my head. When my little brother with the heart condition was fighting for his life, Mom was relieved I was learning music at the conservatory instead of riding back and forth with her to the hospital each day. The alternative to music lessons, she said, "was nothing."

In that moment, with a dental dam stretched across my jaw, it dawned on me that I had no idea what my childhood would have been like without bathing each day in the music of Brahms, Elgar, and Bach. I thought of my siblings, raised in the same home, and the challenges we have all faced. I wouldn't trade places with any of them. While each had their own talents and interests, I was the only one in the thrall of an all-consuming passion, one that transported me on waves of vibration from the chaos in our home.

Perhaps my rigid classical training wasn't solely to blame for the depression, perfectionism, and physical injuries I had endured. Music, instead, might have offered a source of strength and constancy, one that stayed with me through the years as I grew into the journalist, mother, and dabbling musician I am today. Maybe—and the thought startled me in the dentist's chair—music was there for me all along.

This possibility changed everything. Extrapolating from it, I imagined myself savoring the good times playing a tarantella for my first cello teacher, and honoring the memory of performing

Bach to commemorate the fourteen women slain in Montreal. Without the numbing of musical PTSD, I could fully grieve the virtuoso sisters lost in the Air India attack. And maybe, after all this time, I could finally drop the mantle of "failed cellist."

As I lay there, half sedated, waves of sadness engulfed me, along with deep gratitude. For music. For my mother. For the genius of the human brain. After all, it's our brain that weaves layers of meaning from all the chaotic frequencies we hear. As Mendel Kaelen, the neuroscientist, pointed out, "The human mind makes music from sound."

Knowing this, I could receive the gifts of my synapses and fully reclaim my passion for music, even the kind that hurt. I could cherish the burst of wonder I had felt when I first dragged a bow across the cello.

I could feel blessed.

Coda

Do you still play the cello? Will you ever play it again?

I have a feeling people will keep asking me these questions even after reading this book, because there's something irresistible about the cello—its human shape, physicality, and earthy sound. I get it. Part of me still loves it, too.

My cello hasn't left its case, though, and my bow is still in desperate need of rehairing. But beside the couch is a bass dunun drum hewn from yellow cedar fallen in a storm. A friend designed a glass table that allows this drum to be admired from every angle, and on this table, I keep my mbira handy.

I pick up these instruments whenever I get the itch to play. Simply, imperfectly, no strings attached.

Will my cello ever escape from exile? Maybe. When I first started writing this book, I consulted a coach, Marial Shea, who described the process as a transformative experience. Now I know what she meant. Mine started with exploring the parallels between traditional healing strategies and new discoveries in music as medicine. I planned to investigate music through an analytical lens, with no intention of dredging up my feelings. Only through the writing did I recognize another, hidden motivation: to heal myself.

I won't sugarcoat it: The trip down memory lane has been fraught. Every newspaper clipping, journal entry, and artifact from my cello years set off an avalanche of sorrow and regret. I tensed up when I remembered the agony of tendinitis, and wept for days when I revisited the Air India tragedy.

At the same time, reflecting on the history of Western European music and the conservatory system gave me a fresh understanding of the cello training I went through. My teachers didn't mean to be cruel or harsh. They saw talent in me and did everything they could to give me a shot at success. I have forgiven them.

It is harder to forgive myself. As a young person, I saw the world through "everything is against me" eyes. Now I see how much I had going for me and how many people tried to help me shine. It saddens me to realize how often I got in my own way through anxiety, fear, and lack of faith in myself.

On a happier note, looking back has reminded me of my tactile connection to the cello, which is truly unlike any other instrument I've played. At times I've felt the urge to hold it again. Even so, finding my way back to my first instrument is not a burning need. If it happens, it will be a gentle rapprochement, a spontaneous moment that may not lead to more.

I don't dwell on it, because I know that even if I never touch the cello again, I'll always find joy in new rhythms and songs. The pianist Sergei Rachmaninoff had a very different destiny from mine, but I am fond of his motto: "Music is enough for a whole lifetime, but a lifetime is not enough for music."

ACKNOWLEDGMENTS

IF ROB SANDERS, publisher of Greystone Books, hadn't invited me to discuss book ideas over coffee, *Wired for Music* wouldn't exist. He said my initial pitches sounded saleable enough, but then he looked me in the eye. "Do you have anything else?" I told him about my nerdy interest in music and health, and that I'd just turned down a spot in a graduate program in ethnomusicology. Then I mentioned my checkered past with the cello. His eyes brightened. "That's the one."

Many thanks to Greystone for publishing my first book, to editor Lucy Kenward for her structural advice, copy editor Erin Parker for her lovely manner, and editorial director Jennifer Croll, proofreader Jennifer Stewart, and text designer Belle Wuthrich for going above and beyond in the final stages.

Credit also goes to my agent extraordinaire, Martha Webb, for believing in this book through its many detours.

I am extremely grateful for friends and mentors who gave crucial feedback along the way: Angie Abdou, Dominic Ali, Susie Berg, Galya Chatterton, Sylvia Coleman, Elee Kraljii Gardiner, James Glave, Sarah Hampson, Jane Henry, Jillian Horton, Caitlin Kelly, Carol Murray, Susan Olding, Sue Robins, Gary Ross, Amy Kiara Ruth, Eric Unmacht, and Jennifer Van Evra. I owe you big-time.

Colleagues at *The Globe and Mail* taught me a great deal about writing and science journalism. In particular, Dakshana Bascaramurty, Zosia Bielski, Ian Brown, Wency Leung, Hayley Mick, Chris Nuttall-Smith, Paul Taylor, and Carol Toller. Sinclair Stewart approved my leave for a family gap year (sorry for taking a buyout at the end!) and Jana Pruden gave me excellent advice on structure. Early in this book project, a pivotal conversation with Marsha Lederman rescued it from the brink.

All my life I've been blessed to have teachers of music, formal and informal. Special thanks to Navaro Franco, David Hutchenreuther, and Alexandra Jai. A big "Obrigada!" to Kleber Magrão and other members of Brazil's mangue scene. And as they say in Zimbabwe, "Tatenda!" to the musicians and scholars who shared their knowledge of mbira music and Chivanhu culture with me: Musekiwa Chingodza, Jonathan Goredema, Kurai Mubaiwa, Ambuya Mugwagwa, Fradreck Mujuru, Patience Munjeri, Chiedza Mutamba, Florence Mutamba, Moyo Rainos Mutamba, Alois Mutinhiri, Caution Shonhai, Joyce Warikandwa, and the lovely people at Ubuntu Learning Village. Much appreciation, too, to Erica Azim of mbira.org and Jennifer Kyker at Eastman School of Music.

As a layperson, I couldn't have written this book without the many researchers and scientists who agreed to speak with me or review technical passages, including Bernd Brabec de Mori, University of Innsbruck; Wade Davis, National Geographic Society; Jessica Grahn, Western University; Ethan Hein, New York University; Henkjan Honing, University of Amsterdam; Mendel Kaelan, Wavepaths; Costas Karageorghis, Brunel University London; Anton Killin, Australian National University; Kevin Kirkland, Capilano University; Samuel Mehr, Harvard University; Moyo Rainos Mutamba, Ubuntu Learning Village; Andrew Newberg, Thomas Jefferson University; David J. Rothenberg, Case Western Reserve University; Michael Thaut, University of Toronto; Laurel Trainor, McMaster University; and Robert Zatorre, Montreal Neurological Institute and McGill University. (Any errors are entirely my own.)

I can't imagine finishing a book without friends and family to commiserate with and celebrate small victories. Eternal gratitude to Ksenia Barton, Marta Becker, Galya Chatterton, Sylvia Coleman, Emily Corse, Sandrine de Finney, Jennifer Van Evra, Hal Wake, Adele Weder, and many others.

At the sentence-by-sentence level, though, no one helped more than my empathic book coach turned dear friend, Marial Shea. From the proposal stage to manuscript delivery, she was like a mother to the writer in me. (If anyone needs hand-holding through the writing process, hire her!)

To my way-cooler-than-me parents, Susan Feindel and Russell Barton, thank you for setting the course for an unconventional life—and for never reproaching me for quitting the cello. Your acceptance helped me heal.

To my beloved husband and son, I know it wasn't easy living with my frazzled side in the final months of this writing project. Scott, you have supported me in every possible way, from paying the bills when I left my *Globe* job to believing in me when I couldn't. I cherish you and all the joys you have brought to my life.

Finally, a big thank-you to every stranger who said they'd like to read a book like mine, from a distinguished music critic to a waiter at the Keg. Every reader is a gift.

NOTES

Introduction

4 **some of the same brain pathways stimulated by chocolate or sex:** Anne J. Blood and Robert J. Zatorre, "Intensely Pleasurable Responses to Music Correlate With Activity in Brain Regions Implicated in Reward and Emotion," *Proceedings of the National Academy of Sciences of the United States of America* 98, no. 20 (September 25, 2001): 11818–23, doi.org/10.1073/pnas.191355898.

"musical anhedonia": "Lack of Joy From Music Linked to Brain Disconnection," The Neuro, January 4, 2017, mcgill.ca/neuro/channels/news/lack-joy-music-linked-brain-disconnection-264862.

Noelia Martínez-Molina et al., "Neural Correlates of Specific Musical Anhedonia," *Proceedings of the National Academy of Sciences of the United States of America* 113, no. 46 (November 15, 2016): E7337–45, doi.org/10.1073/pnas.1611211113.

music fires up the putamen: Jessica A. Grahn, email correspondence with author, February 1, 2022.

Jessica A. Grahn, "The Role of the Basal Ganglia in Beat Perception: Neuroimaging and Neuropsychological Investigations," *Annals of the New York Academy of Sciences* 1169, no. 1 (July 2009): 35–45, doi.org/10.1111/j.1749-6632.2009.04553.x.

"Ian, lift your finger": Adriana Barton, "The Man Inside: A Wounded Policeman's Life on the Frontier of Consciousness," *The Globe and Mail*, August 18, 2018, theglobeandmail.com/canada/article-the-man-inside-a-wounded-policemans-life-on-the-frontier-of/.

6 **"Decade of the Brain," new brain-imaging techniques:** "Decade of the Brain," Library of Congress, accessed January 15, 2022, loc.gov/loc/brain/.

William R. Uttal, "The Two Faces of MRI," *Dana Foundation* (blog), July 1, 2002, dana.org/article/the-two-faces-of-mri/.

rhythm and song fire up important brain regions: Daniel J. Levitin, *This Is Your Brain on Music: The Science of a Human Obsession* (London: Plume Book, 2007), 86–87.

Irigwe people of central Nigeria: Anthony E. Mereni, *Music Therapy: Concept, Scope and Competence* (Lagos, Nigeria: Apex Books Limited, 2004).

Peter O. Ebigbo, "The Mind, the Body, and Society: An African Perspective," *Advances: Journal of the Institute for the Advancement of Health* 3, no. 4 (1986): 45–57.

Tuvan healers: Pat Moffitt Cook and Julian Burger, *Music Healers of Indigenous Cultures: Shaman, Jhankri & Néle*, Second updated and revised edition (Bainbridge Island, WA: Open Ear Press, 2004), 48.

music and the minimally conscious state: Manon Carrière et al., "An Echo of Consciousness: Brain Function During Preferred Music," *Brain Connectivity* 10, no. 7 (September 1, 2020): 385–95, doi.org/10.1089/brain.2020.0744.

Yajuan Hu et al., "Can Music Influence Patients With Disorders of Consciousness? An Event-Related Potential Study," *Frontiers in Neuroscience* 15 (April 9, 2021): 596636, doi.org/10.3389/fnins.2021.596636.

Fabien Perrin et al., "Promoting the Use of Personally Relevant Stimuli for Investigating Patients With Disorders of Consciousness," *Frontiers in Psychology* 6 (2015), frontiersin.org/article/10.3389/fpsyg.2015.01102.

Fabien Perrin, email correspondence with author, February 1, 2022.

music for Second World War veterans: U.S. War Department, "Technical Bulletin 187: Music in Reconditioning in American Service Forces Convalescent and General Hospitals," War Department Technical Bulletin, TB Med 187 (Washington, DC: U.S. War Department, 1945).

M. A. Rorke, "Music and the Wounded of World War II," *Journal of Music Therapy* 33, no. 3 (September 1, 1996): 189–207, doi.org/10.1093/jmt/33.3.189.

7 **Harold Rhodes's "xylette":** Hugh Davies, *Rhodes, Harold*, vol. 1 (Oxford University Press, 2001), doi.org/10.1093/gmo/9781561592630.article.47652.

Alan S. Lenhoff and David E. Robertson, *Classic Keys: Keyboard Sounds That Launched Rock Music* (Denton, TX: University of North Texas Press, 2019), 230.

music in cancer care: Joke Bradt et al., "Music Interventions for Improving Psychological and Physical Outcomes in Cancer Patients," *Cochrane Database of Systematic Reviews*, no. 8 (2016), doi.org/10.1002/14651858.CD006911.pub3.

music and anxiety: Joke Bradt, Cheryl Dileo, and Minjung Shim, "Music Interventions for Preoperative Anxiety," *Cochrane Database of Systematic Reviews*, no. 6 (2013), doi.org/10.1002/14651858.CD006908.pub2.

music and dementia: Ronald Devere, "Music and Dementia: An Overview," *Practical Neurology* (Bryn Mawr Communications, June 2017), practicalneurology.com/articles/2017-june/music-and-dementia-an-overview.

music versus meditation: Kim E. Innes et al., "Effects of Meditation Versus Music Listening on Perceived Stress, Mood, Sleep, and Quality of Life in Adults With Early Memory Loss: A Pilot Randomized Controlled Trial," *Journal of Alzheimer's Disease: JAD* 52, no. 4 (April 8, 2016): 1277–98, doi.org/10.3233/JAD-151106.

"like a hole in my heart": Wade Davis, Zoom interview with author, December 11, 2020.

8 **musical expression as birthright:** J. A. Sloboda, Karen J. Wise, and Isabelle Peretz, "Quantifying Tone Deafness in the General Population," *Annals of the New York Academy of Sciences* 1060, no. 1 (December 1, 2005): 255–61, doi.org/10.1196/annals.1360.018.

Christopher Small, *Musicking: The Meanings of Performing and Listening* (Hanover, NH: University Press of New England, 1998), 8.

some African and Indigenous languages don't have a specific word for "singer" or "musician": Bruno Nettl, *The Study of Ethnomusicology: Thirty-One Issues and Concepts* (University of Illinois Press, 2010), 16–26.

Timothy Rice, *Ethnomusicology: A Very Short Introduction*, Very Short Introductions (New York: Oxford University Press, 2014).

"transformative technology of the mind": Aniruddh Patel and Henkjan Honing, "Music as a Transformative Technology of the Mind: An Update," in *The Origins of Musicality* (Cambridge, MA:MIT Press), 113–26, accessed January 14, 2022, tufts.app.box.com/v/psy-pub-patel-2018-music-trans.

9 **Maslow's "hierarchy of needs":** A. H. Maslow, "A Theory of Human Motivation," *Psychological Review* 50, no. 4 (July 1943): 370–96, doi.org/10.1037/h0054346.

Maslow's "hierarchy of needs" was likely inspired by his experiences living on a Blackfoot Reserve in the summer of 1938: Narcisse Blood and Ryan Heavy Head, "Blackfoot Influence on Abraham Maslow," (University of Montana, October 27, 2007), blackfootdigitallibrary.com/digital/collection/bdl/id/1296/rec/1.

10 **negative effects of social media use:** Brian A. Primack et al., "Social Media Use and Perceived Social Isolation Among Young Adults in the U.S.," *American Journal of Preventive Medicine* 53, no. 1 (July 2017): 1–8, doi.org/10.1016/j.amepre.2017.01.010.

Jessica C. Levenson et al., "The Association Between Social Media Use and Sleep Disturbance Among Young Adults," *Preventive Medicine* 85 (April 2016): 36–41, doi.org/10.1016/j.ypmed.2016.01.001.

Brian A. Primack et al., "Use of Multiple Social Media Platforms and Symptoms of Depression and Anxiety: A Nationally-Representative Study Among U.S. Young Adults," *Computers in Human Behavior* 69 (April 1, 2017): 1–9, doi.org/10.1016/j.chb.2016.11.013.

music and dopamine: Valorie N. Salimpoor et al., "Anatomically Distinct Dopamine Release During Anticipation and Experience of Peak Emotion to Music," *Nature Neuroscience* 14, no. 2 (February 2011): 257–62, doi.org/10.1038/nn.2726.

1: Strings Attached

15 **Yo-Yo Ma's Stradivari cello:** "Yo-Yo Ma on Playing His 1712 'Davidov' Stradivari Cello," *The Strad*, September 28, 2021, thestrad.com/playing-and-teaching/yo-yo-ma-on-playing-his-1712-davidov-stradivari-cello/13689.article.

"Antonio Stradivari, Cremona, 1712, the 'Davidoff,'" *Tarisio* (blog), accessed January 13, 2022, tarisio.com/cozio-archive/property/.

16 **founding of *The Strad* magazine:** "*The Strad*: About Us," *The Strad*, accessed January 13, 2022, thestrad.com/about.

Cleveland Orchestra listed among the top-five orchestras in the world: James R. Oestreich, "Out From Under the Shadow," *The New York Times*, January 26, 1997, sec. Arts, nytimes.com/1997/01/26/arts/out-from-under-the-shadow.html.

19 **treatment for ganglion cysts:** "Ganglion Cysts: The Bible Bump," The Orthopedic Clinic, August 20, 2018, orthotoc.com/ganglion-cysts-the-bible-bump/.

"Ganglion Cyst: Diagnosis and Treatment," Mayo Clinic, accessed January 13, 2022, mayoclinic.org/diseases-conditions/ganglion-cyst/diagnosis-treatment/drc-20351160.

23 **Tchaikovsky, The Beatles, Carnegie Hall:** "Pyotr Ilyich Tchaikovsky," Carnegie Hall, accessed January 13, 2022, carnegiehall.org/About/History/Carnegie-Hall-Icons/Pyotr-Ilyich-Tchaikovsky.

Michael Pollak, "Meeting the Beatles," *The New York Times*, November 12, 2006, sec. New York, nytimes.com/2006/11/12/nyregion/thecity/meeting-the-beatles.html.

25 **"universal language":** Henry Wadsworth Longfellow, *Outre-mer: A Pilgrimage Beyond the Sea* (New York: Harper & Brothers, 1835).

26 **Tsimane people and music:** Josh H. McDermott et al., "Indifference to Dissonance in Native Amazonians Reveals Cultural Variation in Music Perception," *Nature* 535, no. 7613 (July 2016): 547–50, doi.org/10.1038/nature18635.

Congolese hunter-gatherers and music: Hauke Egermann et al., "Music Induces Universal Emotion-Related Psychophysiological Responses: Comparing Canadian Listeners to Congolese Pygmies," *Frontiers in Psychology* 5 (2015), frontiersin.org/article/10.3389/fpsyg.2014.01341.

most of us can identify another culture's dance tune or lullaby: Samuel A. Mehr et al., "Form and Function in Human Song," *Current Biology* 28, no. 3 (February 2018): 356-368.e5, doi.org/10.1016/j.cub.2017.12.042.

we don't share a common musical language: Ethan Hein, "Music Is Not a Universal Language and This Klezmer Song Proves It," *The Ethan Hein Blog* (blog), May 5, 2021, ethanhein.com/wp/2021/music-is-not-a-universal-language-and-this-klezmer-song-proves-it/.

2: The Music Instinct

30 **no such thing as a "music gene" or a "center in the brain that Stevie Wonder has that nobody else does":** WIRED staff, "Music Makes Your Brain Happy," *WIRED* Magazine, August 23, 2006, wired.com/2006/08/music-makes-your-brain-happy/.

"music is as ubiquitous a capacity among humans as is language": "Where It All Began," Penn Today: University of Pennsylvania, January 10, 2008, penntoday.upenn.edu/node/149663.

31 **"auditory cheesecake":** Steven Pinker, *How the Mind Works* (New York: Norton, 2009), 534.

"and the rest of our lifestyle would be virtually unchanged": Pinker, 528.

"From Darwin onward, great minds pondered questions like ...": University of Sydney, *Alfred Hook Lecture: Gary Tomlinson*, 2015, youtube.com/watch?v=A8RAS9gcB8k.

origins of music: Gary Tomlinson, *A Million Years of Music: The Emergence of Human Modernity* (New York: Zone Books, 2015).

32 **"as an essential part of human identity, rivaling speech":** Matthew Franke, "Review: A Million Years of Music by Gary Tomlinson," *MAKE Literary Magazine* (blog), April 17, 2018, makemag.com/review-a-million-years-of-music-by-gary-tomlinson/.

33 **beat perception is "functional at birth":** Winkler et al., "Newborn Infants Detect the Beat in Music," *Proceedings of the National Academy of Sciences* 106, no. 7 (February 17, 2009): 2468–71, doi.org/10.1073/pnas.0809035106.

34 **clocks would eventually swing together, in "odd sympathy":** Christian Huygens, "Correspondance 1664–1665." In Nijhoff, M. (ed.) *Oeuvres complètes de Christiaan Huygens*, vol. 5 (The Hague: La Société Hollandaise des Sciences, 1893).

musical animals: Henkjan Honing, *The Evolving Animal Orchestra: In Search of What Makes Us Musical* (Cambridge, MA: The MIT Press, 2019), 9.

35 **"my jaw hit the floor":** Claudia Dreifus, "Exploring Music's Hold on the Mind," *The New York Times*, May 31, 2010, sec. Science, nytimes.com/2010/06/01/science/01conv.html.

macaques share 93 percent of their DNA with humans: "Analysis of Rhesus Monkey Genome Uncovers Genetic Differences With Humans, Chimps," National Institutes of Health (NIH), April 12, 2007,nih.gov/news-events/news-releases/analysis-rhesus-monkey-genome-uncovers-genetic-differences-humans-chimps.

chimpanzee swaying to a beat, and limited vocal learning: Eva Frederick, "Dancing Chimpanzees May Reveal How Humans Started to Boogie," *Science*, December 23, 2019, science.org/content/article/dancing-chimpanzees-may-reveal-how-humans-started-boogie.

Pedro Tiago Martins and Cedric Boeckx, "Vocal Learning: Beyond the Continuum," *PLOS Biology* 18, no. 3 (March 30, 2020): e3000672, doi.org/10.1371/journal.pbio.3000672.

36 **last common ancestor:** Darren Curnoe, "When Humans Split From the Apes," The Conversation, February 21, 2016, theconversation.com/when-humans-split-from-the-apes-55104.

two colossal skulls of concrete mark the turnoff to the "Cradle of Humankind": Peter Dorfman, "Stone Age Institute Designs Monument at Olduvai Gorge," *Bloom Magazine*, January 3, 2020, magbloom.com/2020/01/stone-age-institute-designs-monument-at-olduvai-gorge/.

Oldupai/Olduvai Gorge and Laetoli Footprints: "Ngorongoro Conservation Area," *Africa Geographic* (blog), July 22, 2021, africageographic.com/stories/ngorongoro-conservation-area/.

"Mystery Solved: Footprints From Site A at Laetoli, Tanzania, Are From Early Humans, Not Bears," *The Leakey Foundation* (blog), December 1, 2021, leakeyfoundation.org/mystery-solved-footprints-from-site-a-at-laetoli-tanzania-are-from-early-humans-not-bears/.

37 **massive growth spurt in the hominin brain:** "History Module: The Expansion of the Hominid Brain," The Brain, McGill University, accessed January 28, 2022, thebrain.mcgill.ca/flash/capsules/histoire_bleu04.html.

Note: the "hominid" family includes apes and orangutans, whereas "hominin" refers to *Homo sapiens* and our closest extinct relatives.

theories about origins of beat perception: Anton Killin, "Musicality and the Evolution of Mind, Mimesis, and Entrainment," *Biology & Philosophy* 31, no. 3 (May 2016): 421–34, doi.org/10.1007/s10539-016-9519-1.

Tomlinson, *A Million Years of Music*.

It took focus—and several hundred hours of practice—to learn to knap a hand-axe: Dietrich Stout et al., "Cognitive Demands of Lower Paleolithic Toolmaking," *PLOS ONE* 10, no. 4 (April 15, 2015): e0121804, doi.org/10.1371/journal.pone.0121804.

Dietrich Stout, email correspondence with author, January 28, 2022.

38 **fire and human evolution:** Nicholas Mott, "What Makes Us Human? Cooking, Study Says," *National Geographic*, October 26, 2012, nationalgeographic.com/animals/article/121026-human-cooking-evolution-raw-food-health-science.

Chris Organ et al., "Phylogenetic Rate Shifts in Feeding Time During the Evolution of *Homo*," *Proceedings of the National Academy of Sciences* 108, no. 35 (August 30, 2011): 14555–59, doi.org/10.1073/pnas.1107806108.

Anton Killin, "The Origins of Music: Evidence, Theory, and Prospects," *Music & Science* 1 (January 1, 2018): 205920431775197, doi.org/10.1177/2059204317751971.

39 **"I was his roadie and manager":** Michael Fles, phone interview with author, April 26, 2018.

Christopher Tree: Pieter Van Deusen and Christopher Tree, *Christopher Tree*, documentary short, 1967.

40 **"Woodstock, the '60s, and some hitchhikers":** Kevin Kirkland, *Celebrating the Beginnings of the First Canadian Music Therapy Program*, 2017, youtube.com/watch?v=aayB7LdvRFO.

41 **embodied, imitative, and emotional aspects of human cognition:** Dylan van der Schyff and Andrea Schiavio, "Evolutionary Musicology Meets Embodied Cognition: Biocultural Coevolution and the Enactive Origins of Human Musicality," *Frontiers in Neuroscience* 11 (2017), frontiersin.org/article/10.3389/fnins.2017.00519.

"fillet steaks were sliced from the spine, and the bones were smashed to get out the marrow": "The Evolution of Man," BBC, September 17, 2014, bbc.co.uk/sn/prehistoric_life/human/human_evolution/first_europeans1.shtml.

nonverbal communication at Boxgrove site: University of Sydney, *Alfred Hook Lecture: Gary Tomlinson*.

anatomical changes in vocal and auditory systems: Bart de Boer, "Loss of Air Sacs Improved Hominin Speech Abilities," *Journal of Human Evolution* 62, no. 1 (January 2012): 1–6, doi.org/10.1016/j.jhevol.2011.07.007.

I. Martínez et al., "Auditory Capacities in Middle Pleistocene Humans From the Sierra de Atapuerca in Spain," *Proceedings of the National Academy of Sciences* 101, no. 27 (July 6, 2004): 9976–81, doi.org/10.1073/pnas.0403595101.

42 **pitch perception in humans and non-human primates:** Levitin, *This Is Your Brain on Music*, 42–43.

Olivier Joly et al., "A Perceptual Pitch Boundary in a Non-Human Primate," *Frontiers in Psychology* 5 (2014), frontiersin.org/article/10.3389/fpsyg.2014.00998.

rise of *Homo sapiens*: Céline M. Vidal et al., "Age of the Oldest Known *Homo Sapiens* From Eastern Africa," *Nature* 601, no. 7894 (January 27, 2022): 579–83, doi.org/10.1038/s41586-021-04275-8.

to make music, early humans had to learn to combine smaller components: University of Sydney, *Alfred Hook Lecture: Gary Tomlinson*.

Tomlinson, *A Million Years of Music*.

we don't need our language system to process music: Xuanyi Chen et al., "The Human Language System Does Not Support Music Processing," preprint (Neuroscience, June 1, 2021), doi.org/10.1101/2021.06.01.446439.

43 **ice-age flutes:** Killin, "The Origins of Music."

Tom Service, "The Ice-Age Flute That Can Play the Star-Spangled Banner," *The Guardian*, February 15, 2013, sec. Music, theguardian.com/music/2013/feb/15/ice-age-Flute.

biological purposes of music: Levitin, *This Is Your Brain on Music*, 247–67.

babies will calm down to lullabies in any language: Constance M. Bainbridge et al., "Infants Relax in Response to Unfamiliar Foreign Lullabies," *Nature*

Human Behaviour 5, no. 2(February 2021): 256–64, doi.org/10.1038/s41562-020-00963-z.

unborn babies respond to the contours of melodies . . . "musicality precedes both music and language": Honing, *The Evolving Animal Orchestra*, 123–24.

44 **"pop-sci myth that refuses to die":** Samuel Mehr, email correspondence with author, January 27, 2022.

the human auditory cortex can rewire itself to process touch as well as sound: Walter Neary, "Brains of Deaf People Rewire to 'Hear' Music," *University of Washington News* (blog), November 27, 2001, washington.edu/news/2001/11/27/brains-of-deaf-people-rewire-to-hear-music/.

"like a huge ear": Robert Everett-Green, "Dame Evelyn Glennie, the Deaf Percussionist Who Listens With Her Whole Body," *The Globe and Mail*, March 1, 2011, theglobeandmail.com/arts/music/dame-evelyn-glennie-the-deaf-percussionist-who-listens-with-her-whole-body/article568725/.

In Western industrial societies, up to a quarter of adults will describe themselves as unmusical: L. L. Cuddy, "Musical Difficulties Are Rare: A Study of 'Tone Deafness' Among University Students," *Annals of the New York Academy of Sciences* 1060, no. 1 (December 1, 2005): 311–24, doi.org/10.1196/annals.1360.026.

Sally Bodkin-Allen, Nicola Swain, and Susan West, "'It's Not That Bad Singing With Other People': The Effect of a Single Outreach on Singing Attitudes and Confidence in Adults," *Australian Journal of Music Education* 52 (December 6, 2019): 95–105.

J. A. Sloboda, Karen J. Wise, and Isabelle Peretz, "Quantifying Tone Deafness in the General Population," *Annals of the New York Academy of Sciences* 1060, no. 1 (December 1, 2005): 255–61, doi.org/10.1196/annals.1360.018.

45 **infant formula, breastfeeding:** Emily E. Stevens, Thelma E. Patrick, and Rita Pickler, "A History of Infant Feeding," *Journal of Perinatal Education* 18, no. 2 (January 1, 2009): 32–39, doi.org/10.1624/105812409x426314.

Andrew J. Schuman, "A Concise History of Infant Formula (Twists and Turns Included)," *Contemporary Pediatrics*, 2003, contemporarypediatrics.com/view/concise-history-infant-formula-twists-and-turns-included.

S. J. Fomon, "Reflections on Infant Feeding in the 1970s and 1980s," *The American Journal of Clinical Nutrition* 46, no. 1 (July 1,1987): 171–82, doi.org/10.1093/ajcn/46.1.171.

"central for human well-being": van der Schyff and Schiavio, "Evolutionary Musicology Meets Embodied Cognition."

46 **looking at pictures of nature, compared to urban scenes, improved people's scores by 20 percent:** Marc G. Berman, John Jonides, and Stephen Kaplan, "The Cognitive Benefits of Interacting With Nature," *Psychological Science* 19, no. 12 (December 2008): 1207–12, doi.org/10.1111/j.1467-9280.2008.02225.x.

listening to a few minutes of slow-paced rhythms calms breathing and pulse: Martina de Witte et al., "Effects of Music Interventions on Stress-Related Outcomes: A Systematic Review and Two Meta-Analyses," *Health Psychology Review* 14, no. 2 (April 2, 2020): 294–324, doi.org/10.1080/17437199.2019.1627897.

"what it means to be human and alive": "Wade Davis: Light at the Edge of the World," Ideas, CBC Radio, November 26, 2018, cbc.ca/radio/ideas/wade-davis-light-at-the-edge-of-the-world-1.4499962.

3: Groove, Interrupted

50 **"People think of Brazil as this dreamy land of rainforests...":** Kleber Magrão, in-person conversations with author, Recife, Brazil, January 15–29, 2000.

51 **details on Recife, Brazil, "Kings of Congo," and abolishment of slavery in Brazil:** "Recife History—Culture, Religion and Lifestyle in Recife," Recife.com, accessed January 26, 2022, recife.com/v/history/.

Schuyler Whelden and Juliana Cantarelli Vita, "Maracatu de Baque Virado," Massa: Brazilian Music and Culture, May 11, 2021, essefoimassa.com/episodes/tag/kings+of+congo.

The Brazilian Report, "Slavery in Brazil," Wilson Center, May 13, 2020, wilsoncenter.org/blog-post/slavery-brazil.

53 **"There's this strange phenomenon that people are afraid to express themselves musically...":** CopperCat Band, *Daniel Levitin Interview—From Boomer Bands Documentary*, 2018, youtube.com/watch?v=M_CWY2XR6RO.

amusia: Isabelle Peretz and Dominique T. Vuvan, "Prevalence of Congenital Amusia," *European Journal of Human Genetics* 25, no. 5 (May 2017): 625–30, doi.org/10.1038/ejhg.2017.15.

Krista L. Hyde and Isabelle Peretz, "Brains That Are Out of Tune but In Time," *Psychological Science* 15, no. 5 (May 2004): 356–60, doi.org/10.1111/j.0956-7976.2004.00683.x.

self-perceptions of "tone-deafness": Cuddy, "Musical Difficulties Are Rare."

Bodkin-Allen, Swain, and West, "'It's Not That Bad Singing With Other People.'"

Sloboda, Wise, and Peretz, "Quantifying Tone Deafness in the General Population."

"beat deaf": Jessica Phillips-Silver et al., "Born to Dance but Beat Deaf: A New Form of Congenital Amusia," *Neuropsychologia* 49, no. 5 (April 2011): 961–69, doi.org/10.1016/j.neuropsychologia.2011.02.002.

A lack of beat perception is even rarer than tonal amusia: Jessica Phillips-Silver et al., "Amusic Does Not Mean Unmusical: Beat Perception and Synchronization Ability Despite Pitch Deafness," *Cognitive Neuropsychology* 30, no. 5 (July 2013): 311–31, doi.org/10.1080/02643294.2013.863183.

Honing, *The Evolving Animal Orchestra*, 50.

54 **painful emotions and "selective mutism for singing":** Nicola Swain and Sally Bodkin-Allen, "Can't Sing? Won't Sing? Aotearoa/New Zealand 'Tone-Deaf' Early Childhood Teachers' Musical Beliefs," *British Journal of Music Education* 31, no. 3 (November 2014): 245–63, doi.org/10.1017/S0265051714000278.

Bodkin-Allen, Swain, and West, "'It's Not That Bad Singing With Other People.'"

"fatal blow" to musical self-image: Steven M. Demorest, "Stop Obsessing Over Talent—Everyone Can Sing," The Conversation, March 16, 2017, theconversation.com/stop-obsessing-over-talent-everyone-can-sing-74047.

personal, social, and cultural beliefs about singing: Steven M. Demorest, Jamey Kelley, and Peter Q. Pfordresher, "Singing Ability, Musical Self-Concept, and Future Music Participation," *Journal of Research in Music Education* 64, no. 4 (January 2017): 405–20, doi.org/10.1177/0022429416680096.

Heather Nelson Shouldice, "An Investigation of Musical Ability Beliefs and Self-Concept Among Fourth-Grade Students in the United States," *International Journal of Music Education* 38, no. 4 (November 2020): 525–36, doi.org/10.1177/0255761420914667.

Colleen Whidden, "Understanding Complex Influences Affecting Participation in Singing," in *Proceedings of the Phenomenon of Singing International Symposium*, ed. T. Rheynish (St. John's: Memorial University Press, 2009), 142–56, semanticscholar.org/paper/Understanding-Complex-Influences-Affecting-in-Whidden/1bf7520c61dd0b036faa85f8c7866e61d6bac817.

55 **music in ancient Greece:** Penelope Murray and Peter Wilson, *Music and the Muses: The Culture of Mousike in the Classical Athenian City* (Oxford University Press, 2004), 23, 32. doi.org/10.1093/acprof:oso/9780199242399.001.0001.

Donald Jay Grout and Claude V. Palisca, *A History of Western Music*, 4th ed. (New York: Norton, 1988).

Pythagoras and music: Peregrine Horden, *Music as Medicine: The History of Music Therapy Since Antiquity* (Routledge, 2016), 55–56.

The Editors of Encyclopaedia Britannica, "Pythagoras: Biography, Philosophy, and Facts," Britannica.com, accessed January 22, 2022, britannica.com/biography/Pythagoras.

Carl Huffman, "Pythagoras," in *Stanford Encyclopedia of Philosophy*, ed. Edward N. Zalta, 2018 ed. (Metaphysics Research Lab, Stanford University, 2018), plato.stanford.edu/archives/win2018/entries/pythagoras/.

Plato and the "harmony of the spheres": Horden, *Music as Medicine*, 58, 64.

music in ancient Rome: Grout and Palisca, *A History of Western Music*, 23–24.

Mark Cartwright, "Roman Games, Chariot Races & Spectacle," *World History Encyclopedia*, 2013, worldhistory.org/article/635/roman-games-chariot-races--spectacle/.

Bacchus beckoned worshippers to drink wine and let loose to the "howlings" of drums: "Bacchanalia," *New World Encyclopedia*, accessed January 22, 2022, newworldencyclopedia.org/entry/Bacchanalia.

Livy. Books 38–39 With an English Translation (London: William Heinemann, Ltd., 1936), bk. 39, ch. 8.

Sarah Limoges, "Expansionism or Fear: The Underlying Reasons for the Bacchanalia Affair of 186 B.C.," *Hirundo, the McGill Journal of Classical Studies*, vol. 7 (September 2008): 77–94.

56 **"devious spells of syncopated tunes . . .":** *The Fathers of the Church: A New Translation*, vol. 23 (The Catholic University of America Press Inc., 1953), 130.

While a harp was permitted to accompany hymns at home . . . : Grout and Palisca, *A History of Western Music*, 25.

"better suited to beasts": *The Fathers of the Church*, 130–31.

"diabolical choruses": Grout and Palisca, *A History of Western Music*, 34.

"sinning criminally": Grout and Palisca, 36.

57 **Gregorian chant, the official music of the Catholic Church:** Grout and Palisca, 67.

Devil's tritone: Don Michael Randel, ed., *The Harvard Dictionary of Music*, 4th ed. (Cambridge, MA: Belknap Press of Harvard University Press, 2003).

"You can read into that a theological ban in the guise of a technical ban": Finlo Rohrer, "The Devil's Music," *BBC News*, April 28, 2006, http://news.bbc.co.uk/2/hi/uk_news/magazine/4952646.stm.

58 **the singing of harmonies "must not give empty pleasure to the ear...":** Grout and Palisca, *A History of Western Music*, 320.

U.S. classical music consumption in 2020: Nielsen Music/MRC Data, "Mid-Year Report: U.S. 2020," 2020, 33.

pop stars who got their start in church choirs: Maddy Shaw Roberts, "11 Pop Stars You'd Never Believe Used to Be Choristers," Classic FM, December 24, 2018, classicfm.com/discover-music/pop-stars-choristers-church-choirs/.

59 **I found write-ups on my first teachers in the Canadian Encyclopedia:** Gilles Potvin, "André Mignault," The Canadian Encyclopedia, May 9, 2007, thecanadianencyclopedia.ca/en/article/andre-mignault-emc.

Denis Allaire, "Josèphe Colle," The Canadian Encyclopedia, July 30, 2007, thecanadianencyclopedia.ca/en/article/josephe-colle-emc.

60 **history of classical music conservatories:** The Editors of Encyclopaedia Britannica, "Conservatory: Musical Institution," Britannica.com, accessed January 22, 2022, britannica.com/art/conservatory-musical-institution.

"The Four Conservatories of Naples: The First Music Conservatories," Neapolitan Music Society, accessed January 22, 2022, neapolitanmusicsociety.org/history.html.

"Le conservatoire de musique et d'art dramatique du québec, près de 80 ans d'histoire," Le Conservatoire de musique et d'art dramatique du Québec, accessed January 22, 2022, conservatoire.gouv.qc.ca/fr/a-propos/historique/.

62 **standard music lessons start at the wrong level of abstraction:** Ethan Hein, "Musical Simples," *The Ethan Hein Blog* (blog), July 7, 2015, ethanhein.com/wp/2015/musical-simples/.

63 **the waltz reached "peak cultural salience" at least 150 years ago:** Ethan Hein, "Teaching Whiteness in Music Class," *The Ethan Hein Blog* (blog), May 8, 2018, ethanhein.com/wp/2018/teaching-whiteness-in-music-class/.

in a 2017 survey, more than 90 percent of elementary and high school music teachers in the U.S. were white: Wendy K. Matthews and Karen Koner, "A Survey of Elementary and Secondary Music Educators' Professional Background, Teaching Responsibilities and Job Satisfaction in the United States," *Research & Issues in Music Education* 13, no. 1 (2017): art. 2, commons.lib.jmu.edu/rime/vol13/iss1/2.

"the aesthetic preferences of Western European aristocrats of the eighteenth and nineteenth centuries": Ethan Hein, "Developing an Intro-Level Music Theory Course," *The Ethan Hein Blog* (blog), June 27, 2019, ethanhein.com/wp/2019/developing-an-intro-level-music-theory-course/.

hip-hop "vanishingly unusual" in American music classrooms: Hein, "Teaching Whiteness in Music Class."

hip-hop has a lot to offer: Adam J. Kruse, "Being Hip-Hop: Beyond Skills and Songs," *General Music Today* 30, no. 1 (October 2016): 53–58, doi.org/10.1177/1048371316658931.

64 **a manual aimed at encouraging kids to use digital tools:** Will Kuhn and Ethan Hein, *Electronic Music School: A Contemporary Approach to Teaching Musical Creativity* (New York: Oxford University Press, 2021).

classical music could become more relevant in today's multicultural classrooms . . . : Hein, "Teaching Whiteness in Music Class."

"no one cares if it would sound good in the morning": Patrick Lewis Wilkie, in-person interview with author, Vancouver, Canada, November 27, 2019.

Billie Eilish: Mathias Rosenzweig, "Meet Billie Eilish, Pop's Next It Girl," *Vogue*, August 9, 2016, vogue.com/article/billie-eilish-pops-next-it-girl.

Rolling Stone, *Billie Eilish and Finneas Break Down Her Hit Song "Bad Guy,"* 2019, youtube.com/watch?v=KPX2-EMfdbg.

65 **national survey of U.S. high schools found that just a quarter of graduating students had enrolled in a single music course during any of their four years in high school:** Kenneth Elpus and Carlos R. Abril, "Who Enrolls in High School Music? A National Profile of U.S. Students, 2009–2013," *Journal of Research in Music Education* 67, no. 3 (October 2019): 323–38, doi.org/10.1177/0022429419862837.

"should we blame the kids for voting with their feet?": Ethan Hein, "Everyone Can and Should Be Making Music," *The Ethan Hein Blog* (blog), February 17, 2014, ethanhein.com/wp/2014/everyone-can-and-should-be-making-music/.

Americans listen to music, on average, roughly twenty-seven hours a week:

Weekly Time Spent Listening to Music in the United States From 2015 to 2019," Statista, accessed January 17, 2022, statista.com/statistics/828195/time-spent-music/.

"it's become more of a background thing, kind of like auditory wallpaper": CopperCat Band, *Daniel Levitin Interview—From Boomer Bands Documentary.*

66 **nearly 85 percent of the recordings we now consume come from streaming services:** Shawn Knight, "Music Industry Revenues Increased 27 Percent in the First Half of 2021, RIIA Report Finds," *TechSpot*, September 13, 2021, techspot.com/news/91228-music-industry-revenues-increased-27-percent-first-half.html.

Streamed audio files do not preserve all the audio information from the original recordings: Robert Triggs, "The Important Difference Between Audio and File Compression," SoundGuys, September 15, 2021, soundguys.com/audio-compression-explained-29148/.

"People can't get what they need from listening to music anymore . . . ": Neil Young, *Waging Heavy Peace: A Hippie Dream* (New York: Plume, 2013).

lossless streaming: "Apple Music Announces Spatial Audio and Lossless Audio," Apple Newsroom (Canada), May 17, 2021, apple.com/ca/newsroom/2021/05/apple-music-announces-spatial-audio-and-lossless-audio/.

"Five Things to Know About Spotify HiFi," Spotify, February 22, 2021, newsroom.spotify.com/2021-02-22/five-things-to-know-about-spotify-hifi/.

"like you've never heard it before": "TIDAL: High Fidelity Music Streaming," TIDAL—High Fidelity Music Streaming, accessed January 25, 2022, tidal.com/.

in 2020, vinyl records accounted for just 5 percent of U.S. music revenues: "U.S. Sales Database," RIAA, accessed January 17, 2022, riaa.com/u-s-sales-database/.

they go for the fashion scene, VIP pool parties, and ice-cream tacos . . . : Hillary Hoffower, "15 Unbelievable Facts Show Just How Much It Costs to Attend Coachella, From $430 Tickets to $9,500 Campsites," *Business Insider*, accessed January 25, 2022, businessinsider.com/coachella-cost-tickets-parking-travel-food-festival-fashion-2019-4.

"The music is almost incidental to the experience": Ross Gerber, "How the Music Industry Is Putting Itself Out of Business," *Forbes*, May 3, 2017, forbes.com/sites/greatspeculations/2017/05/03/how-the-music-industry-is-putting-itself-out-of-business/.

67 **"shadow" musicians:** Julia Cameron, *The Artist's Way: A Spiritual Path to Higher Creativity*, 25th anniversary edition (New York: TarcherPeregree, 2016).

"You've just invented a new form of torture": Kathryn L. Braun, "'In a Competition Full of Hamburgers, You're a Steak:' American Idol and the Role of Reality Television in the Maintenance of Our Egos," *Kaleidoscope* 7, article 6, 7.

the cruelty is part of the point: Ujala Sehgal, "New 'Nice' American Idol's Ratings Sink Without Simon Cowell," *Business Insider*, accessed January 25, 2022, businessinsider.com/new-nice-american-idol-ratings-sink-without-simon-cowell-2011-1.

68 **"mirror neurons":** Honing, *The Evolving Animal Orchestra*, 39.

4: Mood Music

69 **"music releases mood-enhancing chemical in the brain":** Sonya McGilchrist, "Music 'Releases Mood-Enhancing Chemical in the Brain,'" *BBC News*, January 9, 2011, sec. Health, bbc.com/news/health-12135590.

70 **Air India bombing:** Public Safety Canada, "Remembering Air India Flight 182," Government of Canada, December 21, 2018, publicsafety.gc.ca/cnt/ntnl-scrt/cntr-trrrsm/r-nd-flght-182/index-en.aspx.

72 **Elgar Cello Concerto and Jacqueline du Pré:** Sarah Kirkup, "Jacqueline du Pré and Elgar's Cello Concerto," Gramophone, May 25, 2021, gramophone.co.uk/features/article/jacqueline-du-pre-and-elgar-s-cello-concerto.

"Elgar Cello Concerto: Jacqueline du Pré: The Concerto's Consummate Interpreter?," Elgar.org, accessed January 18, 2022, elgar.org/3cello-b.htm.

73 **"Their mood state is extremely unpleasant . . . ":** Kerry Sheridan, "Study Finds Surprising Benefit to Listening to Sad Music," WUSF Public Media, May 1, 2019, wusfnews.wusf.usf.edu/university-beat/2019-05-01/study-finds-surprising-benefit-to-listening-to-sad-music.

people with clinical depression showed a strong preference for somber music:

Sunkyung Yoon et al., "Why Do Depressed People Prefer Sad Music?," *Emotion* 20, no. 4 (2020): 613–24, doi.org/10.1037/emo0000573.

the psychology of sad music: Jaak Panksepp, "The Emotional Sources of Chills Induced by Music," *Music Perception* 13, no. 2 (December 1, 1995): 171–207, doi.org/10.2307/40285693.

Kazuma Mori and Makoto Iwanaga, "Two Types of Peak Emotional Responses to Music: The Psychophysiology of Chills and Tears," *Scientific Reports* 7, no. 1 (April 7, 2017): 46063, doi.org/10.1038/srep46063.

74 **"cry a little and then feel relieved, and move on":** Annemieke J. M. Van den Tol and Jane Edwards, "Exploring a Rationale for Choosing to Listen to Sad Music When Feeling Sad," *Psychology of Music* 41, no. 4 (July 2013): 440–65, doi.org/10.1177/0305735611430433.

poignant songs invite us to savor emotions such as sadness and melancholy: Tuomas Eerola et al., "An Integrative Review of the Enjoyment of Sadness Associated With Music," *Physics of Life Reviews* 25 (August 1, 2018): 100–121, doi.org/10.1016/j.plrev.2017.11.016.

"he was inconsolable": Jennifer Van Evra, personal communication with author, November 6, 2019.

75 **"completely abstract and profoundly emotional . . .":** Oliver Sacks, *Musicophilia: Tales of Music and the Brain* (Toronto: Vintage Canada, 2008), 329.

music at the nineteenth-century psychiatric institution Illenau: Horden, *Music as Medicine*, 346.

"I had chills down my spine. I had goose bumps": Dr. Robert Zatorre, *From Perception to Pleasure: How Music Changes the Brain*, TEDxHECMontréal, 2018, ted.com/talks/dr_robert_zatorre_from_perception_to_pleasure_how_music_changes_the_brain.

76 **music and dopamine:** Robert Zatorre, phone interview with author, July 30, 2021.

Salimpoor et al., "Anatomically Distinct Dopamine Release During Anticipation and Experience of Peak Emotion to Music."

the "Kim Kardashian of molecules": Vaughan Bell, "The Unsexy Truth About Dopamine," *The Observer*, February 3, 2013, sec. Science, theguardian.com/science/2013/feb/03/dopamine-the-unsexy-truth.

endogenous opioids: J. Jacob, "Endogenous Morphines and Pain," *Agents and Actions* 11, no. 6–7 (December 1981): 634–36, doi.org/10.1007/BF01978772.

Eduardo E. Benarroch, "Endogenous Opioid Systems: Current Concepts and Clinical Correlations," *Neurology* 79, no. 8 (August 21, 2012): 807–14, doi.org/10.1212/WNL.0b013e3182662098.

77 **music under the influence of dopamine-blocking and enhancing drugs:** Laura Ferreri et al., "Dopamine Modulates the Reward Experiences Elicited by Music," *Proceedings of the National Academy of Sciences* 116, no. 9 (February 26, 2019): 3793–98, doi.org/10.1073/pnas.1811878116.

Zatorre and colleagues repeated the experiment with an opioid-enhancing drug: Robert Zatorre, phone interview with author, July 30, 2021.

hits of bliss from what neuroscientists call "aesthetic" or "cognitive rewards": Wolfram Schultz, "Multiple Reward Signals in the Brain," *Nature Reviews Neuroscience* 1, no. 3 (December 2000): 199–207, doi.org/10.1038/35044563.

lack of pleasure a hallmark of depression: "Depression (Major Depressive Disorder): Symptoms and Causes," Mayo Clinic, accessed January 19, 2022, mayoclinic.org/diseases-conditions/depression/symptoms-causes/syc-20356007.

78 **"Anxiety . . . because I have it":** Liliana Moreno, in-person conversation with author, Vancouver, Canada, January 26, 2020.

sedative side effects: "Diazepam (Oral Route) Side Effects," Mayo Clinic, accessed January 20, 2022, mayoclinic.org/drugs-supplements/diazepam-oral-route/side-effects/drg-20072333?p=1.

music for preoperative anxiety: P. Berbel, J. Moix, and S. Quintana, "Music Versus Diazepam to Reduce Preoperative Anxiety: A Randomized Controlled Clinical Trial," *Revista Española de Anestesiología y Reanimacíon* 54, no. 6 (2007): 355–58.

Joke Bradt, Cheryl Dileo, and Minjung Shim, "Music Interventions for Preoperative Anxiety," ed. Cochrane Anaesthesia, Critical and Emergency Care Group, *Cochrane Database of Systematic Reviews*, June 6, 2013, doi.org/10.1002/14651858.CD006908.pub2.

79 **anxiety and stress:** "Anxiety Disorders: Symptoms and Causes," Mayo Clinic, accessed January 19, 2022, mayoclinic.org/diseases-conditions/anxiety/symptoms-causes/syc-20350961.

"How Stress Affects Your Body and Behavior," Mayo Clinic, accessed January 19, 2022, mayoclinic.org/healthy-lifestyle/stress-management/in-depth/stress-symptoms/art-20050987.

Gail Innis, "Understanding Cortisol, the Stress Hormone," Michigan State University, November 27, 2018, canr.msu.edu/news/understanding_cortisol_the_stress_hormone.

"moderate tranquilising" effects of music: de Witte et al., "Effects of Music Interventions on Stress-Related Outcomes."

80 **dopamine increases most when we enjoy the music:** Salimpoor et al., "Anatomically Distinct Dopamine Release During Anticipation and Experience of Peak Emotion to Music."

listening preference and brain connectivity: R. W. Wilkins et al., "Network Science and the Effects of Music Preference on Functional Brain Connectivity: From Beethoven to Eminem," *Scientific Reports* 4, no. 1 (December 17, 2014): 6130, doi.org/10.1038/srep06130.

Tarantella and spider bites: Horden, *Music as Medicine*, 275, 281, 283, 286, 295.

"Spider Bites and Venoms," The Australian Museum, accessed January 19, 2022, australian.museum/learn/animals/spiders/spider-bites-and-venoms.

Douglas J. Lanska, "The Dancing Manias: Psychogenic Illness as a Social Phenomenon," in *Frontiers of Neurology and Neuroscience*, ed. J. Bogousslavsky, vol. 42 (S. Karger AG, 2018), 132–41, doi.org/10.1159/000475719.

81 **musicians wandering the fields in southern Italy:** Samuel Pepys, *The Diary of Samuel Pepys M.A. F.R.S.*, Reproduction of the original (Frankfurt am Main: Outlook Verlag, 2018), 15.

"highly convincing" evidence that music improves symptoms of depression and quality of life: Daniel Leubner and Thilo Hinterberger, "Reviewing the Effectiveness of Music Interventions in Treating Depression," *Frontiers in Psychology* 8 (July 7, 2017): 1109, doi.org/10.3389/fpsyg.2017.01109.

"a funeral" in the brain: Emily Dickinson and R. W. Franklin, *The Poems of Emily Dickinson*, Variorum ed. (Cambridge, MA: Belknap Press of Harvard University Press, 1998).

music and depression: "Depression: Key Facts," World Health Organization, accessed January 19, 2022, who.int/news-room/fact-sheets/detail/depression.

up to two-thirds of clinically diagnosed people may improve with anti-depressants and talk therapy: Khalid Saad Al-Harbi, "Treatment-Resistant Depression: Therapeutic Trends, Challenges, and Future Directions," *Patient Preference and Adherence* 6 (May 1, 2012): 369–88, doi.org/10.2147/PPA.S29716.

music therapy offered an extra boost compared to standard treatments alone: Sonja Aalbers et al., "Music Therapy for Depression," ed. Cochrane Common Mental Disorders Group, *Cochrane Database of Systematic Reviews* 2017, no. 11 (November 16, 2017), doi.org/10.1002/14651858.CD004517.pub3.

82 **music therapy training, certified profession:** "A Career in Music Therapy," American Music Therapy Association, accessed January 19, 2022, musictherapy.org/.

In a study of cancer patients with low mood . . . : Joke Bradt et al., "The Impact of Music Therapy Versus Music Medicine on Psychological Outcomes and Pain in Cancer Patients: A Mixed Methods Study," *Supportive Care in Cancer* 23, no. 5 (May 2015): 1261–71, doi.org/10.1007/s00520-014-2478-7.

"music medicine" (definition): Bradt et al. (2015).

the brain's endogenous opioid system may also be directly involved in regulating mood . . . : Marta Peciña et al., "Endogenous Opioid System Dysregulation in Depression: Implications for New Therapeutic Approaches," *Molecular Psychiatry* 24, no. 4 (April 2019): 576–87, doi.org/10.1038/s41380-018-0117-2.

A 2020 analysis reported a "significant reduction" in depression symptoms: Qishou Tang et al., "Effects of Music Therapy on Depression: A Meta-Analysis of Randomized Controlled Trials," ed. Sukru Torun, *PLOS ONE* 15, no. 11 (November 18, 2020): e0240862, doi.org/10.1371/journal.pone.0240862.

Studies have used everything from European classical to Indian ragas, Irish folk to reggae, and lullabies to rock: Leubner and Hinterberger, "Reviewing the Effectiveness of Music Interventions in Treating Depression."

83 **The more we like the music, the better our chances of experiencing a mild, depression-lifting euphoria:** Leubner and Hinterberger.

Zatorre (2021).

music has been described as an "emerging treatment option" for mood disorders . . . : Leubner and Hinterberger.

"The capacity of art, music, and dance to circumvent the speechlessness that comes with terror . . . ": Bessel van der Kolk, *The Body Keeps the Score: Brain, Mind, and Body in the Healing of Trauma* (New York: Penguin Publishing Group, 2015), 243.

music and trauma in Rwanda: Anna Leach, "Exporting Trauma: Can the Talking Cure Do More Harm Than Good?," *The Guardian*, February 5, 2015, sec. Working in development, theguardian.com/global-development-professionals-network/2015/feb/05/mental-health-aid-western-talking-cure-harm-good-humanitarian-anthropologist.

Andrew Solomon, "Notes on an Exorcism," *The Moth*, October 29, 2008, themoth.org/stories/notes-on-an-exorcism.

"Rwanda Youth Music," *Musicians Without Borders* (blog), accessed January 19, 2022, musicianswithoutborders.org/programs/rwanda-youth-music/.

84 **trust in a practitioner can be enough to trigger a placebo effect:** Michelle Dossett, phone interview with author, September 18, 2016.

traumatic experiences of Carter Viss and Leila Viss: Leila Viss, phone interview with author, March 30, 2021.

Joe Capozzi, "Hit by a Boat in the Ocean, He Watched His Arm Fall Off. Now Carter Viss Tells His Tale of Survival," *Palm Beach Post*, October 16, 2020, palmbeachpost.com/in-depth/news/local/2020/10/16/florida-snorkeler-fights-water-safety-after-hit-boat-breakers-reef/3635469001/.

87 **"After silence that which comes nearest to expressing the inexpressible is music":** Aldous Leonard Huxley, "The Rest Is Silence," in *Music at Night and Other Essays* (New York: Doubleday Doran and Company, Inc., 1931), 17, archive.org/details/in.ernet.dli.2015.166397/page/n21/mode/2up.

88 **"We were actually pushing their immune systems in the wrong direction . . .":** Bary Bittman, phone interview with author, September 12, 2016.

music and immune markers: B. Bittman et al., "Composite Effects of Group Drumming Music Therapy on Modulation of Neuroendocrine-Immune Parameters in Normal Subjects," *Alternative Therapies in Health and Medicine* 7, no. 2 (2001): 62–73.

5: A Musician's Brain

91 ***Voyager* probes and the Golden Record:** "*Voyager*: The Golden Record," NASA Jet Propulsion Laboratory, accessed January 18, 2022, voyager.jpl.nasa.gov/Golden-record/.

The lead researcher, Frances Rauscher . . . : Joseph B. Verrengia, "Lab Rats Listen to Mozart, Become Maze-Busters," *Los Angeles Times*, November 8, 1998, latimes.com/archives/la-xpm-1998-nov-08-mn-40406-story.html.

"one of the most profound and mature" of all Mozart's compositions: G. L. Shaw, *Keeping Mozart in Mind* (Amsterdam; Boston: Elsevier, 2004).

92 **Rauscher described her study in a two-page letter in the journal *Nature*:** Frances H. Rauscher, Gordon L. Shaw, and Catherine N. Ky, "Music and Spatial Task Performance," *Nature* 365, no. 6447 (October 14, 1993): 611, doi.org/10.1038/365611a0.

the "Mozart effect": J. S. Jenkins, "The Mozart Effect," *Journal of the Royal Society of Medicine* 94, no. 4 (April 2001): 170–72, doi.org/10.1177/014107680109400404.

Adrian Bangerter and Chip Heath, "The Mozart Effect: Tracking the Evolution of a Scientific Legend," *British Journal of Social Psychology* 43, no. 4 (December 2004): 605–23, doi.org/10.1348/0144666042565353.

Rauscher and her coauthor merely hypothesized . . . : Elaine Woo, "Gordon Shaw, 72; Linked Music to Thinking," *Los Angeles Times*, May 1, 2005, latimes.com/archives/la-xpm-2005-may-01-me-shaw1-story.html.

"You can never control what the marketers will do . . .": Robert Lee Hotz, "Study Finds That Mozart Music Makes You Smarter: Science," *Los Angeles Times*, October 14, 1993, latimes.com/archives/la-xpm-1993-10-14-mn-45497-story.html.

"No one questions . . .": Kevin Sack, "Georgia's Governor Seeks Musical Start for Babies," *The New York Times*, January 15, 1998, sec. U.S., nytimes.com/1998/01/15/us/georgia-s-governor-seeks-musical-start-for-babies.html.

It was clear the Mozart myth didn't hold up: Kenneth M. Steele, "The 'Mozart Effect': An Example of the Scientific Method in Operation," *Psychology Teacher Network*, December 2001, 2–5, appstate. edu/~steelekm/documents/Steele2001_ptnpdf.

93 **children and adults with extensive music training tend to outperform non-musicians on tests of working memory, attention, and executive functioning . . . :** Jennifer Zuk et al., "Behavioral and Neural Correlates of Executive Functioning in Musicians and Non-Musicians," ed. Amanda Bruce, *PLOS ONE* 9, no. 6 (June 17, 2014): e99868, doi.org/10.1371/journal.pone.0099868.

Brenda Hanna-Pladdy and Byron Gajewski, "Recent and Past Musical Activity Predicts Cognitive Aging Variability: Direct Comparison With General Lifestyle Activities," *Frontiers in Human Neuroscience* 6 (2012), doi.org/10.3389/fnhum.2012.00198.

David Medina and Paulo Barraza, "Efficiency of Attentional Networks in Musicians and Non-Musicians," *Heliyon* 5, no. 3 (March 2019): e01315, doi.org/10.1016/j.heliyon.2019.e01315.

Rafael Román-Caballero, Elisa Martín-Arévalo, and Juan Lupiáñez, "Attentional Networks Functioning and Vigilance in Expert Musicians and Non-Musicians," *Psychological Research* 85, no. 3 (April 2021): 1121–35, doi.org/10.1007/s00426-020-01323-2.

Assal Habibi et al., "Music Training and Child Development: A Review of Recent Findings From a Longitudinal Study," *Annals of the New York Academy of Sciences* 1423, no. 1 (July 2018): 73–81, doi.org/10.1111/nyas.13606.

Learning an instrument at a young age has been linked to stronger auditory processing, emotional perception, and "stick-to-itiveness"...: Adam T. Tierney, Jennifer Krizman, and Nina Kraus, "Music Training Alters the Course of Adolescent Auditory Development," *Proceedings of the National Academy of Sciences* 112, no. 32 (August 11, 2015): 10062–67, doi.org/10.1073/pnas.1505114112.

Sylvain Moreno et al., "Musical Training Influences Linguistic Abilities in 8-Year-Old Children: More Evidence for Brain Plasticity," *Cerebral Cortex* 19, no. 3 (March 2009): 712–23, doi.org/10.1093/cercor/bhn120.

Christine Nussbaum and Stefan R. Schweinberger, "Links Between Musicality and Vocal Emotion Perception," *Emotion Review* 13, no. 3 (July 2021): 211–24, doi.org/10.1177/17540739211022803.

Dana L. Strait et al., "Musical Experience and Neural Efficiency: Effects of Training on Subcortical Processing of Vocal Expressions of Emotion," *European Journal of Neuroscience* 29, no. 3 (February 2009): 661–68, doi.org/10.1111/j.1460-9568.2009.06617.x.

Franziska Degé, Claudia Kubicek, and Gudrun Schwarzer, "Music Lessons and Intelligence: A Relation Mediated by Executive Functions," *Music Perception* 29, no. 2 (December 1, 2011): 195–201, doi.org/10.1525/mp.2011.29.2.195.

the "marshmallow test": Walter Mischel, Ebbe B. Ebbesen, and Antonette Raskoff Zeiss, "Cognitive and Attentional Mechanisms in Delay of Gratification," *Journal of Personality and Social Psychology* 21, no. 2 (1972): 204–18, doi.org/10.1037/h0032198.

Jessica McCrory Calarco, "Why Rich Kids Are So Good at the Marshmallow Test," *The Atlantic*, June 1, 2018, theatlantic.com/family/archive/2018/06/marshmallow-test/561779/.

music, linguistic skills, reading, academic success: Moreno et al. Kathleen A. Corrigall and Laurel J. Trainor, "Associations Between Length of Music Training and Reading Skills in Children," *Music Perception* 29, no. 2 (December 1, 2011): 147–55, doi.org/10.1525/mp.2011.29.2.147.

Adrian Hille and Jürgen Schupp, "How Learning a Musical Instrument Affects the Development of Skills," *Economics of Education Review* 44 (February 2015): 56–82, doi.org/10.1016/j.econedurev.2014.10.007.

Martin Guhn, Scott D. Emerson, and Peter Gouzouasis, "A Population-Level Analysis of Associations Between School Music Participation and Academic

Achievement," *Journal of Educational Psychology* 112, no. 2 (February 2020): 328, doi.org/10.1037/edu0000431.

95 **no relationship between music lessons and enhanced cognitive skills or academic performance:** Giovanni Sala and Fernand Gobet, "Cognitive and Academic Benefits of Music Training With Children: A Multilevel Meta-Analysis," *Memory & Cognition* 48, no. 8 (November 2020): 1429–41, doi.org/10.3758/s13421-020-01060-2.

96 **The team described efforts to enhance academic skills through music training as "pointless":** "Music Training May Not Make Children Smarter After All," ScienceDaily, July 28, 2020, sciencedaily.com/releases/2020/07/200728201550.htm.

personality traits and other pre-existing differences likely explain the link between music training and high grades: Kathleen A. Corrigall, E. Glenn Schellenberg, and Nicole M. Misura, "Music Training, Cognition, and Personality," *Frontiers in Psychology* 4 (2013), doi.org/10.3389/fpsyg.2013.00222.

"anatomists . . . would recognize the brain of a professional musician without a moment's hesitation": Sacks, *Musicophilia*, 100.

first MRI scans of a musician's brain: Gottfried Schlaug, *Music and the Brain (Podcasts): Library of Congress*, 2010, loc.gov/podcasts/musicandthebrain/podcast_schlaug.html.

structure of a musician's brain: G. Schlaug et al., "Increased Corpus Callosum Size in Musicians," *Neuropsychologia* 33, no. 8 (August 1995): 1047–55, doi.org/10.1016/0028-3932(95)00045-5.

Christian Gaser and Gottfried Schlaug, "Brain Structures Differ Between Musicians and Non-Musicians," *Journal of Neuroscience* 23, no. 27 (October 8, 2003): 9240–45, doi.org/10.1523/JNEUROSCI.23-27-09240.2003.

Sibylle C. Herholz and Robert J. Zatorre, "Musical Training as a Framework for Brain Plasticity: Behavior, Function, and Structure," *Neuron* 76, no. 3 (November 8, 2012): 486–502, doi.org/10.1016/j.neuron.2012.10.011.

a musician "is basically an auditory-motor athlete": Schlaug, *Music and the Brain (Podcasts): Library of Congress*.

97 **a musician's brain comes from nurture, not nature:** Krista L. Hyde et al., "The Effects of Musical Training on Structural Brain Development: A Longitudinal Study," *Annals of the New York Academy of Sciences* 1169, no. 1 (July 2009): 182–86, doi.org/10.1111/j.1749-6632.2009.04852.x.

Örjan de Manzano and Fredrik Ullén, "Same Genes, Different Brains: Neuroanatomical Differences Between Mono-zygotic Twins Discordant for Musical Training," *Cerebral Cortex* 28, no. 1 (January 1, 2018): 387–94, doi.org/10.1093/cercor/bhx299.

Adrian Imfeld et al., "White Matter Plasticity in the Corticospinal Tract of Musicians: A Diffusion Tensor Imaging Study," *NeuroImage* 46, no. 3 (July 2009): 600–607, doi.org/10.1016/j.neuroimage.2009.02.025.

music wires the brain in specific ways, depending on the instrument we play: Gaser and Schlaug, "Brain Structures Differ Between Musicians and Non-Musicians."

Lara Schlaffke et al., "Boom Chack Boom—A Multimethod Investigation of Motor Inhibition in Professional Drummers," *Brain and Behavior* 10, no. 1 (January 2020), doi.org/10.1002/brb3.1490.

Thomas Elbert et al., "Increased Cortical Representation of the Fingers of the Left Hand in String Players," *Science* 270, no. 5234 (November 1, 1995): 305–7, doi.org/10.1126/science.270.5234.305.

beatboxers show functional brain changes: Saloni Krishnan et al., "Beatboxers and Guitarists Engage Sensorimotor Regions Selectively When Listening to the Instruments They Can Play," *Cerebral Cortex* 28, no. 11 (November 1, 2018): 4063–79, doi.org/10.1093/cercor/bhy208.

98 **"developmental window" for a musician's brain:** Christopher J. Steele et al., "Early Musical Training and White-Matter Plasticity in the Corpus Callosum: Evidence for a Sensitive Period," *Journal of Neuroscience* 33, no. 3 (January 16, 2013): 1282–90, doi.org/10.1523/JNEUROSCI.3578-12.2013.

Jennifer Bailey and Virginia Penhune, "The Relationship Between the Age of Onset of Musical Training and Rhythm Synchronization Performance: Validation of Sensitive Period Effects," *Frontiers in Neuroscience* 7 (2013), frontiersin.org/article/10.3389/fnins.2013.00227.

Hanna-Pladdy and Gajewski, "Recent and Past Musical Activity Predicts Cognitive Aging Variability."

synaptic pruning: Irwin Feinberg, "Why Is Synaptic Pruning Important for the Developing Brain?," *Scientific American*, May 1, 2017, scientificamerican .com/article/why-is-synaptic-pruning-important-for-the-developing-brain/.

99 **music training plays a vital role in child development:** Anita Collins, *The Music Advantage: How Music Helps Your Child Develop, Learn, and Thrive* (New York: TarcherPerigee, 2021).

"you get it wrong more than you get it right . . .": Melbourne Symphony Orchestra, *Melbourne Music Summit Keynote: Dr. Anita Collins: The Ups and Downs of Music Advocacy*, 2021, youtube.com/watch?v=pzkPA-llq4M.

"I taught myself how to play the guitar, I taught myself how to play the drums, and I kind of fake doing both": Steve Appleford, "Dave Grohl Drums Up Probot," *Rolling Stone* (blog), February 6, 2004, rollingstone.com/music /music-news/dave-grohl-drums-up-probot-175443/.

The Off Camera Show, *Dave Grohl Proves You Don't Need Lessons to Rock*, 2013, youtube.com/watch?v=ixo-MQou6SA.

David Bowie: Isaac Guzmán, "In Appreciation: David Bowie, 1947 to 2016," *Time* Magazine, January 14, 2016, time.com/magazine/us/4180259 /january-25th-2016-vol-187-no-2-u-s/.

100 **Jimi Hendrix:** Sean Michaels, "Was Jimi Hendrix's Ambidexterity the Key to His Virtuosity?," *The Guardian*, February 25, 2010, sec. Music, theguardian .com/music/2010/feb/25/jimi-hendrix-ambidexterity-virtuosity.

self-teaching and less restrictive musical environments "tend to enhance creativity": Peter D. MacIntyre and Gillian K. Potter, "Music Motivation and the Effect of Writing Music: A Comparison of Pianists and Guitarists," *Psychology of Music* 42, no. 3 (2013): 403–19, doi.org/10.1177/0305735613477180.

efforts to make classical music training more natural and child-friendly: Alfred Garson and Emily-Jane Orford, "Suzuki Method," *The Canadian Encyclopedia*, February 7, 2006, thecanadianencyclopedia.ca/en/article /suzuki-method-emc.

Judit Eniko Szanto, "Singing Technique for Young Children in the Kodály Music Classroom: A Narrative Inquiry" (University of Calgary, September 24, 2021), hdl.handle.net/1880/114002.

"Our Curriculum: Music for Young Children," Music for Young Children, accessed January 17, 2022, myc.com/about-our-curriculum/.

102 **Albert Einstein and music:** George Sylvester Viereck, "What Life Means to Einstein: An Interview by George Sylvester Viereck," *The Saturday Evening Post*, October 26, 1929.

Mitch Waldrop, "Inside Einstein's Love Affair With 'Lina'—His Cherished Violin," *National Geographic*, February 3, 2017, nationalgeographic.com/adventure/article/einstein-genius-violin-music-physics-science.

Albert Einstein, *The Collected Papers of Albert Einstein*, ed. Anna Beck, trans. Peter Havas (Princeton, NJ: Princeton University Press, 1987), xxi.

Alice Calaprice, ed., *The Ultimate Quotable Einstein* (Princeton, NJ: Princeton University Press, 2010), 239, press.princeton.edu/books/hardcover/9780691138176/the-ultimate-quotable-einstein.

103 **"constantly search for new harmonies and transitions of his own invention":** Einstein, *The Collected Papers of Albert Einstein*, xxi.

"music helps him when he is thinking about his theories . . . ": Abraham Pais, *"Subtle Is the Lord": The Science and the Life of Albert Einstein* (Oxford University Press, 2005), 301.

"Einstein in Debut as Violinist Here; Lewisohn Ballroom Filled for Concert to Aid His Friends in Berlin," *The New York Times*, January 18, 1934.

"there was hardly one whose feeling and understanding for good music was deeper than Einstein's": János Plesch, *János, the Story of a Doctor* (New York: A.A. Wynn, 1949), 214.

104 **"reinforces your confidence in the ability to create":** Joanne Lipman, "Is Music the Key to Success?," *The New York Times*, October 12, 2013, sec. Opinion, nytimes.com/2013/10/13/opinion/sunday/is-music-the-key-to-success.html.

Nobel winners often moonlight as musicians: Robert Root-Bernstein et al., "Arts Foster Scientific Success: Avocations of Nobel, National Academy, Royal Society, and Sigma Xi Members," *Journal of Psychology of Science and Technology* 1, no. 2 (October 1, 2008): 51–63, doi.org/10.1891/1939-7054.1.2.51.

"intricate" and "beautiful" code of life: "The Symphony of Science," The Nobel Prize, March 22, 2019, nobelprize.org/symphony-of-science/.

Nobel prize–winning musicians: Natalie Angier, "Frances Arnold Turns Microbes Into Living Factories," *The New York Times*, May 28, 2019, sec. Science, nytimes.com/2019/05/28/science/frances-arnold-caltech-evolution.html.

"Richard Feynman's Beloved Bongo Drums," Sotheby's, accessed January 17, 2022, sothebys.com/buy/7fab421a-16d9-4d44-80e9-1baf76295fa6/lots/6626c4f6-69d8-4862-8499-20c21c652993.

"The Symphony of Science."

research on problem-solving emphasizes the value of cross-pollinating our pursuits: Vasanth Sarathy, "Real World Problem-Solving," *Frontiers in Human Neuroscience* 12 (2018), frontiersin.org/article/10.3389/fnhum.2018.00261.

music as a "flexibility primer": Devon E. Hinton and Laurence J. Kirmayer, "The Flexibility Hypothesis of Healing," *Culture, Medicine, and Psychiatry* 41, no. 1 (March 2017): 3–34, doi.org/10.1007/s11013-016-9493-8.

105 **"and it sounded horrible . . . I was wracked by guilt . . . ":** Verrengia, "Lab Rats Listen to Mozart, Become Maze-Busters."

perfectionism: Tyler Pia et al., "Perfectionism and Prospective Near-Term Suicidal Thoughts and Behaviors: The Mediation of Fear of Humiliation and Suicide Crisis Syndrome," *International Journal of Environmental Research and Public Health* 17, no. 4 (February 2020): 1424, doi.org/10.3390/ijerph17041424.

6: More Than Meets the Ear

109 **George Harrison and *Call of the Valley*:** Nyay Bhushan, "George Harrison's Life Was Transformed by India, Says Olivia Harrison," *The Hollywood Reporter* (blog), October 22, 2011, hollywoodreporter.com/news/general-news/george-olivia-harrison-mumbai-film-festival-252161/.

stereo headphones invented in 1958: "A Brief History of Headphones," LSTN Sound Co., October 6, 2020, lstnsound.com/blogs/main/a-brief-history-of-headphones.

"Everyone knows what headphones sound like today...": Matt Alt, "The Walkman, Forty Years On," *The New Yorker*, June 29, 2020, newyorker.com/culture/cultural-comment/the-walkman-forty-years-on.

110 **he imagined a future where machines delivered data with the same "under-the-skin intimacy" of the new music player:** Bruce Headlam, "Origins; Walkman Sounded Bell for Cyberspace," *The New York Times*, July 29, 1999, sec. Technology, nytimes.com/1999/07/29/technology/origins-walkman-sounded-bell-for-cyberspace.html.

"The Sony Walkman has done more to change human perception than any virtual reality gadget": Michael Bull, *Sounding Out the City: Personal Stereos and the Management of Everyday Life* (Bloomsbury Academic, 2000), 1, doi.org/10.5040/9781474215541.

"I do still think it's true": William Gibson, email correspondence with author, January 17, 2021.

history of the Sony Walkman: Alt, "The Walkman, Forty Years On."

Meaghan Haire, "A Brief History of the Walkman," *Time* Magazine, January 7, 2009, content.time.com/time/nation/article/0,8599,1907884,00.html.

Gibson, email correspondence with author.

Meira Gebel, "The Walkman Just Turned 40—Here's How Listening to Music Has Changed Over the Years," *Business Insider*, July 1, 2019, businessinsider.com/history-listening-to-music-recorded-walkman-2019-6.

Tom Zito, "Stepping to the Stereo Strut," *Washington Post*, May 12, 1981, washingtonpost.com/archive/lifestyle/1981/05/12/stepping-to-the-stereo-strut/c810a6d9-c054-4b2b-b150-db330cdd08a6/.

111 **launch of iPod, iPhone:** "Apple Presents iPod," Apple Newsroom, October 23, 2001, apple.com/newsroom/2001/10/23Apple-Presents-iPod/.

"Apple Reinvents the Phone With iPhone," Apple Newsroom, January 9, 2007, apple.com/newsroom/2007/01/09Apple-Reinvents-the-Phone-with-iPhone/.

music streaming data: Felix Richter, "Infographic: The Streaming Takeover," Statista Infographics, accessed January 17, 2022, statista.com/chart/8836/streaming-proportion-of-us-music-revenue/.

Knight, "Music Industry Revenues Increased 27 Percent in the First Half of 2021, RIAA Report Finds."

"What is your favourite hype song?": Jana G. Pruden, @jana_pruden, "What Is Your Favourite Hype Song?," Twitter, February 27, 2020.

112 **"It was rap that got my head in the right place":** Emily Selleck, "Barack Obama Perfectly Recites Eminem's 'Lose Yourself' & Fans Are Here for It," *Hollywood Life* (blog), December 11, 2020, hollywoodlife.com/2020/12/11/barack-obama-eminem-lose-yourself-video/.

"It helps me to relax and get into my own little world.": Nicole Puglise, "What Is Michael Phelps Listening to on His Trademark Olympics Headphones?," *The Guardian*, August 8, 2016, sec. Sport, theguardian.com/sport/2016/aug/08/michael-phelps-headphones-music-swimming-olympics-rio.

Rio Olympics: Tiare Dunlap, "Rio 2016: Michael Phelps Reveals Pre-Race Music to PEOPLE," PEOPLE.com, December 2, 2020, people.com/sports/rio-2016-michael-phelps-reveals-pre-race-music-to-people/.

loud and upbeat music before a competition: Peter C. Terry et al., "Effects of Music in Exercise and Sport: A Meta-Analytic Review," *Psychological Bulletin* 146, no. 2 (February 2020): 91–117, doi.org/10.1037/bul0000216.

Decoding Superhuman, *Regulating Performance With Music With Dr. Costas Karageorghis*, Decoding Superhuman (June 25, 2019), youtube.com/watch?v=uBt7xxcvtDI.

influence of music on "power" behaviors: Dennis Y. Hsu et al., "The Music of Power: Perceptual and Behavioral Consequences of Powerful Music," *Social Psychological and Personality Science* 6, no. 1 (January 2015): 75–83, doi.org/10.1177/1948550614542345.

"It appears that listening to music for three minutes can be enough to—snap!—transform the psyche": "Power Jams: Northwestern Magazine," Northwestern, Winter 2014, northwestern.edu/magazine/winter2014/campuslife/power-jams-bass-heavy-playlist-makes-you-stronger.

Mahotella Queens: Qhama Dayile, "With Only One Original Band Member Left, Mahotella Queens Are Still Going Strong," News24, December 2, 2021, news24.com/drum/celebs/news/with-only-one-original-band-member-left-mahotella-queens-are-still-going-strong-20211202.

most people can pick out a love ballad, dance tune, lullaby, or healing song from a culture unknown to them: Samuel A. Mehr et al., "Form and Function in Human Song," *Current Biology* 28, no. 3(February 5, 2018): 356–68.e5, doi.org/10.1016/j.cub.2017.12.042.

Samuel A. Mehr et al., "Universality and Diversity in Human Song," *Science* 366, no. 6468 (November 22, 2019): eaax0868, doi.org/10.1126/science.aax0868.

114 **"people are reliably rating songs that are actually healing songs":** Ed Yong, "A Study Suggests That People Can Hear Universal Traits in Music," *The Atlantic*, January 25, 2018, theatlantic.com/science/archive/2018/01/the-search-for-universal-qualities-in-music-heats-up/551447/.

"A room that is filled with the vibrations of music...": Erin Despard and Richard Despard, "A Bereft Gardener Turns to Music to Soothe," *The Tyee* (November 6, 2020), thetyee.ca/Culture/2020/11/06/Bereft-Gardener-Music-Soothe/.

"You feel actually touched...": *USA Today*, "Yo-Yo Ma Talks COVID, Hope and Anti-Asian Hate," usatoday.com, April 8, 2021.

115 **Haile Gebrselassie, Dorian Yates, and music, sport, and exercise:** Decoding Superhuman, *Regulating Performance With Music With Dr. Costas Karageorghis.*

Costas I. Karageorghis, *Applying Music in Exercise and Sport* (Champaign, IL; London, UK: Human Kinetics, 2017), 14.

"To some extent, the heavy metal on Dorian's Walkman . . .": Decoding Superhuman, *Regulating Performance With Music With Dr. Costas Karageorghis.*

116 **"legal performance-enhancing drug":** Costas I. Karageorghis and David-Lee Priest, "Music in the Exercise Domain: A Review and Synthesis (Part I)," *International Review of Sport and Exercise Psychology* 5, no. 1 (March 2012): 44–66, doi.org/10.1080/1750984x.2011.631026.

music at 120 beats per minute stimulates us to get moving: Hamish G. MacDougall and Steven T. Moore, "Marching to the Beat of the Same Drummer: The Spontaneous Tempo of Human Locomotion," *Journal of Applied Physiology* 99, no. 3 (September 2005): 1164–73, doi.org/10.1152/japplphysiol.00138.2005.

"It is with tracks at this precise tempo that deejays routinely lure people onto a dance floor": Terry et al., "Effects of Music in Exercise and Sport."

Eminem's "Till I Collapse": Edward Cooper, "Spotify Data Proves These Are the 20 Most Popular Workout Tracks, Including Eminem and Travis Scott," *Men's Health*, April 16, 2021, menshealth.com/uk/fitness/lifestyle/a36142847/spotify-workout-songs-pure-gym-study/.

"Workout Song," *Eminem.com* (blog), accessed January 16, 2022, eminem.news?s=workout%20song.

117 **marathons have either restricted or "strongly discouraged" the use of portable music players:** Boston Athletic Association, "B.A.A. Boston Marathon Rules and Policies," 2019, 27.

"NYRR Code of Conduct," New York Road Runners, accessed January 16, 2022, nyrr.org/run/guidelines-and-procedures/code-of-conduct.

slow-paced music can calm us down and prompt our body's hemodynamic response: Costas I. Karageorghis, email correspondence with author, January 24, 2022.

"and gradually bring you down towards a state of homeostasis . . .": Decoding Superhuman, *Regulating Performance With Music With Dr. Costas Karageorghis.*

writing to music: James Parker, "Stephen King on the Creative Process, the State of Fiction, and More," *The Atlantic*, April 12, 2011, theatlantic.com/entertainment/archive/2011/04/stephen-king-on-the-creative-process-the-state-of-fiction-and-more/237023/.

Gabriel García Márquez, *Living to Tell the Tale*, trans. Edith Grossman (New York: Vintage International, 2004).

Samantha Leach, "Stephanie Land Will 'Probably Start Crying' When She Watches 'Maid,'" *Bustle*, October 7, 2021, bustle.com/entertainment/stephanie-land-maid-netflix-show-memoir-quotes.

118 **multitasking is a myth:** Big Think, *Multitasking Is a Myth, and to Attempt It Comes at a Neurobiological Cost*, 2016, youtube.com/watch?v=IM4u-7Z5URk.

"in terms of how it's encoded in the genome": Big Think, *Multitasking Is a Myth, and to Attempt It Comes at a Neurobiological Cost.*

we become even more distracted by music as we age: Sarah Reaves et al., "Turn Off the Music! Music Impairs Visual Associative Memory Performance in Older Adults," *The Gerontologist* 56, no. 3 (June 2016): 569–77, doi.org/10.1093/geront/gnu113.

music and driving: Costas I. Karageorghis et al., "Psychological and Psychophysiological Effects of Music Intensity and Lyrics on Simulated Urban Driving," *Transportation Research Part F: Traffic Psychology and Behaviour* 81 (August 1, 2021): 329–41, doi.org/10.1016/j.trf.2021.05.022.

Costas I. Karageorghis et al., "Psychological, Psychophysiological and Behavioural Effects of Participant-Selected vs. Researcher-Selected Music in Simulated Urban Driving," *Applied Ergonomics* 96 (October 1, 2021): 103436, doi.org/10.1016/j.apergo.2021.103436.

119 **music on the job:** Anthony Wing Kosner, "The Mind at Work: Daniel Levitin on the Secret Life of the Musical Brain," Work in Progress, October 21, 2019, blog.dropbox.com/topics/work-culture/the-mind-at-work-daniel-levitin-on-the-secret-life-of-the-music.

Big Think, *Multitasking Is a Myth, and to Attempt It Comes at a Neurobiological Cost.*

painters and music: Brian Keith Jackson, "Listen to This Playlist and Get in the Heads of Your Favorite Artists," *Vulture*, November 29, 2018, vulture.com/2018/11/artists-on-the-music-they-listen-to-in-the-studio.html.

"Okay, let's try this with some music, but not anything too distracting": "How A Pulitzer-Prize Winning Novelist Thinks About Coffee, Screenplays, and Facebook," *Writing Routines* (blog), October 12, 2017, writingroutines.com/viet-thanh-nguyen/.

120 **immunoglobulins:** "Immunoglobulins," University of Michigan Health, accessed January 16, 2022, uofmhealth.org/health-library/hw41342.

Immunoglobulin A is "particularly responsive to music": Daisy Fancourt, Adam Ockelford, and Abi Belai, "The Psychoneuroimmunological Effects of Music: A Systematic Review and a New Model," *Brain, Behavior, and Immunity* 36 (October 21, 2013), doi.org/10.1016/j.bbi.2013.10.014.

interleukin-6: Athena Chalaris et al., "Interleukin-6 Trans-Signaling and Colonic Cancer Associated With Inflammatory Bowel Disease," *Digestive Diseases (Basel, Switzerland)* 30, no. 5 (2012): 492–99, doi.org/10.1159/000341698.

immune boost from laughter, nature: Mary Payne Bennett and Cecile Lengacher, "Humor and Laughter May Influence Health IV. Humor and Immune Function," *Evidence-Based Complementary and Alternative Medicine: ECAM* 6, no. 2 (June 2009): 159–64, .org/10.1093/ecam/nem149.

Ming Kuo, "How Might Contact With Nature Promote Human Health? Promising Mechanisms and a Possible Central Pathway," *Frontiers in Psychology* 6 (August 25, 2015): 1093, doi.org/10.3389/fpsyg.2015.01093.

anything that calms the stress response might strengthen our defenses against infections: "What Happens When Your Immune System Gets Stressed Out?," Cleveland Clinic, March 1, 2017, health.clevelandclinic.org/what-happens-when-your-immune-system-gets-stressed-out/.

"I thought it was new-age woo-woo": Jeremy Reynolds, "Can Music Boost Your Immune System?," *Pittsburgh Post-Gazette*, April 20, 2020, post-gazette.com/ae/music/2020/04/20/immune-system-music-therapy-health-benefit-study-mind-body-COVID-19/stories/202004150125.

music can relieve pain after surgery—even if people have no memory of hearing it: Hartmuth Nowak et al., "Effect of Therapeutic Suggestions During General Anaesthesia on Postoperative Pain and Opioid Use: Multicentre Randomised Controlled Trial," *BMJ* 371 (December 10, 2020): m4284, doi.org/10.1136/bmj.m4284.

121 **Fiona Mattatall's surgery playlist:** Fiona Mattatall, @FionaMattatall, "Today's Patients' Music Requests for Their Surgeries," Twitter, November 1, 2021, twitter.com/FionaMattatall/status/1455313265513226243.

"the oldest known method for relieving pain": Christine E. Dobek et al., "Music Modulation of Pain Perception and Pain-Related Activity in the Brain, Brain Stem, and Spinal Cord: A Functional Magnetic Resonance Imaging Study," *The Journal of Pain* 15, no. 10 (October 2014): 1057–68, doi.org/10.1016/j.jpain.2014.07.006.

"large pain-reducing effect": Bradt et al., "Music Interventions for Improving Psychological and Physical Outcomes in Cancer Patients."

122 **music and sleep:** Kira V. Jespersen et al., "Music for Insomnia in Adults," *The Cochrane Database of Systematic Reviews*, no. 8 (August 13, 2015): CD010459, doi.org/10.1002/14651858.CD010459.pub2.

Tabitha Trahan et al., "The Music That Helps People Sleep and the Reasons They Believe It Works: A Mixed Methods Analysis of Online Survey Reports," *PLOS ONE* 13, no. 11 (November 14, 2018): e0206531, doi.org/10.1371/journal.pone.0206531.

Summer Jay and Abhinav Singh, "REM Sleep: What It Is and Why It Matters," Sleep Foundation, December 16, 2021, sleepfoundation.org/stages-of-sleep/rem-sleep.

Shane Cochrane, "A Little Night Music," *The Independent*, April 15, 2019, independent.ie/life/health-wellbeing/health-features/a-little-night-music-38009186.html.

"mysterious effects of music that science has yet to address": Gaelen Thomas Dickson and Emery Schubert, "Music on Prescription to Aid Sleep Quality: A Literature Review," *Frontiers in Psychology* 11 (July 28, 2020): 1695, doi.org/10.3389/fpsyg.2020.01695.

binaural beats: Troy Farah, "Binaural Beats: The Auditory Illusion People Claim Can Heal Your Brain," *Discover*, December 11, 2019, discovermagazine.com/mind/binaural-beats-the-auditory-illusion-people-claim-can-heal-your-brain.

Suzannah Weiss, "Do Binaural Beats Produce a Drug-Like Effect?," *Audiofemme* (blog), January 7, 2019, audiofemme.com/high-notes-binaural-beats-digital-drugs/.

"Binaural Doses for Every Imaginable Mood," iDoser.com, accessed January 16, 2022, i-doser.com.

"Binaural Beats," Aspirin Austria, accessed January 16, 2022, aspirin.at/good-vibes/wie-funktionieren-binaurale-beats.

123 **One study poking holes in the hype:** Hector D. Orozco Perez, Guillaume Dumas, and Alexandre Lehmann, "Binaural Beats Through the Auditory Pathway: From Brainstem to Connectivity Patterns," *ENeuro* 7, no. 2 (March 1, 2020), doi.org/10.1523/ENEURO.0232-19.2020.

124 **Meya app and neurolinguistic programming:** "Meya App: The Science of Music Applied to Well-Being," *ABNewswire* (blog), May 3, 2021, abnewswire.com/pressreleases/meya-app-the-science-of-music-applied-to-well-being_541452.html.

Joanna Taylor, "How Music Can Keep You Healthy and Fit," *Evening Standard*, January 16, 2020, standard.co.uk/lifestyle/esmagazine/music-wellness-trends-health-2020-a4335171.html.

Bruce A. Thyer and Monica Pignotti, *Science and Pseudoscience in Social Work Practice* (New York: Springer Publishing Company, 2015).

Sync Project: Chau Tu, "Can Music Be Used as Medicine?," *The Atlantic*, May 7, 2015, theatlantic.com/health/archive/2015/05/can-music-be-used-as-medicine/391820/.

"Sync Project Joins Bose," Sync Project, February 20, 2018, syncproject.co/blog/sync-project-joins-bose. (Website removed as of April 23, 2022.)

"personalized music therapeutics": "Developing Music as Precision Medicine," Sync Project, syncproject.co, accessed January 16, 2022. (Website removed as of April 23, 2022.)

125 **"Spotify already has that nailed down pretty well":** Zatorre, phone interview with author.

neurofeedback and Music to My Brain program: Zatorre.

7: Bad Vibrations

130 **when we hear music we dislike, brain connectivity in our default network drops:** Wilkins et al., "Network Science and the Effects of Music Preference on Functional Brain Connectivity."

earworms: Anne Margriet Euser, Menno Oosterhoff, and Ingrid van Balkom, "Stuck Song Syndrome: Musical Obsessions—When to Look for OCD," *British Journal of General Practice* 66, no. 643 (February 2016): 90–90, doi.org/10.3399/bjgp16x683629.

Victoria J. Williamson et al., "Sticky Tunes: How Do People React to Involuntary Musical Imagery?," *PLOS ONE* 9, no. 1 (January 31, 2014): e86170, doi.org/10.1371/journalpone.0086170.

Kelly Jakubowski et al., "Dissecting an Earworm: Melodic Features and Song Popularity Predict Involuntary Musical Imagery," *Psychology of Aesthetics, Creativity, and the Arts* 11, no. 2 (May 2017): 122–35, doi.org/10.1037/aca0000090.

C. Philip Beaman, Kitty Powell, and Ellie Rapley, "Want to Block Earworms From Conscious Awareness? B(u)y Gum!," *Quarterly Journal of Experimental Psychology* 68, no. 6 (June 2015): 1049–57, doi.org/10.1080/17470218.2015.1034142.

131 **"a magic key, to which the most tightly closed heart opened":** Maria von Trapp, *The Story of the Trapp Family Singers* (Philadelphia, PA: J. B. Lippincott Company, 1949).

"to make music more intensely, more beautifully, more devotedly than ever before": Christopher Buchenholz, "An Artist's Response to Violence," Leonard Bernstein Office, accessed January 16, 2022, leonardbernstein.com/about/humanitarian/an-artists-response-to-violence.

sirens, Battle of Jericho, Waco: Alex Ross, "When Music Is Violence," *The New Yorker*, June 27, 2016, newyorker.com/magazine/2016/07/04/when-music-is-violence.

Danny Gallagher, "Six Songs Used to Torture and Intimidate," *The Wall Street Journal*, July 21, 2009, wsj.com/articles/SB124690400002401641.

132 **"it must also be able to act destructively"**: Ross, "When Music Is Violence."

Adolf Hitler boyhood: "Adolf Hitler: Early Years, 1889–1913," Holocaust Encyclopedia, accessed January 23, 2022, encyclopedia.ushmm.org/content/en/article/adolf-hitler-early-years-1889-1913.

Unknown author, *Portrait Photograph of Adolf Hitler as an 11 Years Old Child*, commons.wikimedia.org/wiki/File:Adolf_Hitler_as_a_child.jpg.

"Here Comes the Bride": Olivia B. Waxman, "How 'Here Comes the Bride' Became a Wedding Music Tradition," *Time* Magazine, January 25, 2016, time.com/5115834/wedding-march-here-comes-the-bride/.

"I was fascinated at once...": Adolf Hitler, *Mein Kampf* (London: Hutchinson, 1969).

Wagner and anti-Semitism: Clive Brown, "Das Judenthum in Der Musik," in *A Portrait of Mendelssohn* (Yale University Press, 2003), doi.org/10.12987/yale/9780300095395.003.0051.

James Loeffler, "Wagner's Anti-Semitism Still Matters: It Helped Define European Anti-Semitism, Especially When It Came to Jewish Music," *The New Republic*, July 5, 2014, newrepublic.com/article/118331/forbidden-music-michael-haas-reviewed-james-loeffler.

Alex Ross, "As If Music Could Do No Harm," *The New Yorker*, August 20, 2014, newyorker.com/culture/cultural-comment/music-harm.

"The reason we perceive the artist Richard Wagner as being great...": Adolf Hitler, "Adolf Hitler: Speech at a NSDAP Meeting in Nuremberg, 3rd November, 1923," in *Sämtliche Aufzeichnungen: 1905–1924*, ed. Eberhard Jäckel and Axel Kuhn, Quellen und Darstellungen zur Zeitgeschichte, Bd. 21 (Stuttgart: Deutsche Verlags-Anstalt, 1980), 1034.

music in the Third Reich: "Music in the Third Reich," Music and the Holocaust, accessed January 16, 2022, holocaustmusic.ort.org/politics-and-propaganda/third-reich/.

133 **"the most German" of all operas**: David B. Dennis, "'The Most German of All German Operas': Die Meistersinger Through the Lens of the Third Reich," in *Wagner's* Meistersinger*: Performance, History, Representation* (Rochester, NY: The University of Rochester Press, 2003), 98–119.

Wagner's "Beckmesser" almost certainly a Jewish caricature: Edward Rothstein, "Beckmesser: Two Villains at a Swipe," *The New York Times*, January 24, 1993, sec. Arts, nytimes.com/1993/01/24/arts/classical-view-beckmesser-two-villains-at-a-swipe.html.

music and censorship in the Third Reich: "Art and Music Under the Third Reich," Music and the Holocaust, accessed January 16, 2022, holocaustmusic.ort.org/politics-and-propaganda/third-reich/entartete-musik/.

radio as "the most influential and important intermediary between a spiritual movement [Nazism] and the nation": Joseph Goebbels, "Der Rundfunk Als Achte Großmacht (Radio as the Eighth Great Power)," in *Signale Der Neuen Zeit: 25 Ausgewählte Reden von Dr. Joseph Goebbels* (Munich: Zentralverlag der NSDAP, 1938), 200, play.google.com/books/reader?id=IUObAAAAMAAJ&pg=GBS.PA200&hl=en_GB.

radios in Nazi Germany: "Nazi Propaganda and Censorship," Holocaust Encyclopedia, accessed January 16, 2022, encyclopedia.ushmm.org/content/en/article/nazi-propaganda-and-censorship.

Maja Adena et al., "Radio and the Rise of the Nazis in Prewar Germany," *The Quarterly Journal of Economics* 130, no. 4 (November 1, 2015): 1885–1939, doi.org/10.1093/qje/qjv030.

"Culture in the Third Reich: Disseminating the Nazi World-view," Holocaust Encyclopedia, accessed January 16, 2022, encyclopedia.ushmm.org/content/en/article/culture-in-the-third-reich-disseminating-the-nazi-worldview.

later models of the "people's receiver" radio had a tiny eagle with a swastika: "Volksempfänger VE 301 Dyn, Philips, in or after 1938," Rijksmuseum, accessed January 16, 2022, rijksmuseum.nl/en/collection/NG-2009-110.

"Those who didn't know the songs were beaten . . .": "The Concentration and Death Camps," Music and the Holocaust, accessed January 16, 2022, holocaustmusic.ort.org/places/camps/.

Primo Levi witnessed prisoners marching to "Rosamunde": Ross, "When Music Is Violence."

"The beating of the big drums and the cymbals reach us continuously and monotonously": "Official Camp Orchestras in Auschwitz," Music and the Holocaust, accessed January 16, 2022, holocaustmusic.ort.org/places/camps/death-camps/auschwitz/camp-orchestras/.

134 **At the end of the war, Hitler clutched his precious Wagner operas—including *The Valkyrie*—in the Berlin bunker where he shot himself:** Howard Taubman, "Original Scores of Wagner Music Feared Lost at Hitler's Bunker," July 27, 1958, sec. 1.

John Cage's "4′33″": Grout and Palisca, *A History of Western Music*, 876.

"humanly organized sound": John Blacking, *How Musical Is Man?*, The Jessie and John Danz Lectures (Seattle: University of Washington Press, 2000), 10.

135 **"Being exposed to music we do not choose means being forced to vibrate according to it . . .":** Felipe Trotta, *Annoying Music in Everyday Life* (New York: Bloomsbury Academic, 2020).

music and loitering: The Associated Press, "California Shop Cranks Classical Music to Dissuade Loitering," *Financial Post*, May 1, 2018, financialpost.com/pmn/business-pmn/california-shop-cranks-classical-music-to-dissuade-loitering.

Annie Karni, "Transit Hubs See Benefits of Baroque in the Background," *The New York Sun*, December 26, 2006, nysun.com/new-york/transit-hubs-see-benefits-of-baroque-in/45689/.

"a lifelong construction" of memories, ideas, feelings, and experiences of belonging: Trotta, *Annoying Music in Everyday Life*.

Plato, the Greek modes, and "bad vibrations": "Musical Mode," *New World Encyclopedia*, accessed January 16, 2022, newworldencyclopedia.org/entry/Musical_mode.

Plato, *The Republic of Plato*, trans. Allan Bloom, 2nd ed. (New York: Basic Books, 1991).

Horden, *Music as Medicine*, 58.

136 **"bad vibrations" in the Enlightenment and onward:** James Kennaway, *Bad Vibrations: The History of the Idea of Music as a Cause of Disease* (Farnham: Ashgate, 2012).

James Kennaway, "From Sensibility to Pathology: The Origins of the Idea of Nervous Music Around 1800," *Journal of the History of Medicine and Allied Sciences* 65, no. 3 (July 1, 2010): 396–426, doi.org/10.1093/jhmas/jrq004.

"So fervid was feminine admiration . . .": Horden, *Music as Medicine*, 342.

Wagner's leitmotifs: Grout and Palisca, *A History of Western Music*, 749.

137 **speculated that *Tristan and Isolde* might drive people insane:** Horden, *Music as Medicine*, 319.

"like the sane person in a community of the mad . . .": Mark Twain, *What Is Man? Collection of Essays on Various Topics by the Famous Creator of Tom Sawyer and Huckleberry Finn.* Originally published in 1906. (BoD–Books on Demand, 2010), 153.

jazz listeners "are actually incapable of distinguishing between good and evil": Anne Shaw Faulkner, "Does Jazz Put the Sin in Syncopation?," *Ladies' Home Journal*, August 1921.

These "back-masked" messages could "turn us into disciples of the Antichrist": John Brackett, "Satan, Subliminals, and Suicide: The Formation and Development of an Antirock Discourse in the United States During the 1980s," *American Music* 36, no. 3 (2018): 271, doi.org/10.5406/americanmusic.36.3.0271.

Eric Nuzum, *Parental Advisory: Music Censorship in America* (New York: Perennial, 2001), 244.

fear of heavy metal music in the 1980s: Kory Grow, "Tipper Gore Reflects on PMRC 30 Years Later," *Rolling Stone*, September 14, 2015, rollingstone.com/politics/politics-news/tipper-gore-reflects-on-pmrc-30-years-later-57862/.

Kory Grow, "Dee Snider on PMRC Hearing: 'I Was a Public Enemy,'" *Rolling Stone* (blog), September 18, 2015, rollingstone.com/music/music-news/dee-snider-on-pmrc-hearing-i-was-a-public-enemy-71205/.

138 **research on death metal as "feel-good" music:** Bill Thompson and Kirk N. Olsen, "Death Metal Is Often Violent and Misogynist Yet It Brings Joy and Empowerment to Fans," The Conversation, February 21, 2018, theconversation.com/death-metal-is-often-violent-and-misogynist-yet-it-brings-joy-and-empowerment-to-fans-91909.

William Forde Thompson, Andrew M. Geeves, and Kirk N. Olsen, "Who Enjoys Listening to Violent Music and Why?," *Psychology of Popular Media Culture* 8, no. 3 (July 2019): 218–32, doi.org/10.1037/ppm0000184.

"perhaps the most important tool of the international neo-Nazi movement to gain revenue and new recruits": Tzvi Fleischer, "Sounds of Hate," *The Review: Australia/Israel & Jewish Affairs Council*, August 2000.

139 **Barney songs as torture: "Trust me, it works . . .":** Adam Piore, "Periscope: PSYOPS: Cruel and Unusual," *Newsweek*, May 19, 2003, newsweek.com/periscope-137213.

Ruhal Ahmed at Guantánamo Bay: David Rose, "Revealed: The Full Story of the Guantanamo Britons," *The Observer*, March 14, 2004, sec. UK news, theguardian.com/uk/2004/mar/14/terrorism.guantanamo.

"Once you accept that you're going to go into the interrogation room and be beaten up . . .": Reprieve.org, *Reprieve: Pull the Plug on Torture Music—Ruhal Ahmed*, 2008, youtube.com/watch?v=_EUllAiFWQC.

Donald Vance at Camp Cropper: Michael Moss, "Former U.S. Detainee in Iraq Recalls Torment," *The New York Times*, December 18, 2006, sec. World, nytimes.com/2006/12/18/world/middleeast/18justice.html.

140 **"began destroying me"**: Suzanne G. Cusick, "'You Are in a Place That Is out of the World . . .': Music in the Detention Camps of the 'Global War on Terror,'" *Journal of the Society for American Music* 2, no. 1 (February 2008), doi.org/10.1017/S1752196308080012.

Amnesty International denounced prolonged and involuntary exposure to loud music as a form of torture: Amnesty International, "Iraq: Torture Not Isolated—Independent Investigations Vital," April 30, 2004, amnesty.org/en/documents/mde14/017/2004/en/.

War crimes, 1949 Geneva Conventions: Office on Genocide Prevention and the Responsibility to Protect, "War Crimes: Background," United Nations, accessed January 15, 2022, un.org/en/genocideprevention/war-crimes.shtml.

Anthony Lewis, "Guantánamo's Long Shadow," *The New York Times*, June 21, 2005, sec. Opinion, nytimes.com/2005/06/21/opinion/guantanamos-long-shadow.html.

musical torture during the War on Terror: David Peisner, "Music as Torture: War Is Loud," *Spin*, November 30, 2006, spin.com/articles/music-torture-war-loud.

Interrogation Log on Detainee 063 at Guantánamo Bay: Pentagon Joint Task Force Guantanamo, "Secret Orcon: Interrogation Log Detainee 063," 84-Page Secret Interrogation Log Obtained by *Time*, n.d., content.time.com/time/2006/log/log.pdf.

Adam Zagorin and Michael Duffy, "Inside the Interrogation of Detainee 063," *Time* Magazine, June 20, 2005, time.com/3624326/inside-the-Interrogation-of-detainee-063/.

al-Qahtani had a history of breakdowns before his imprisonment: Dana Jabri and Shezza Abboushi Dallal, "Al-Qahtani Must Be Released and Guantanamo Closed," January 12, 2021, aljazeera.com/opinions/2021/1/12/qahtani-must-be-released-guantanamo-closed.

141 **"no long-lasting effect"**: BBC, "Sesame Street Breaks Iraqi POWs," *BBC News*, May 20, 2003, news.bbc.co.uk/2/hi/middle_east/3042907.stm.

"Music is used to make the detainee aware that he has no control over what's going on in any of his senses . . .": Alan Connor, "Torture Chamber Music," July 10, 2008, news.bbc.co.uk/2/hi/7495175.stm.

"makes your brain short a fuse": Dan Viau, "Cinema Remembered: Clockwork Orange and the 'Singing in the Rain' Moment," *That Moment In* (blog), January 14, 2016, thatmomentin.com/cinema-remembered-clockwork-orange-and-the-singing-in-the-rain-moment/.

our brains entrain to repetitive musical rhythms: Sylvie Nozaradan, "Exploring How Musical Rhythm Entrains Brain Activity With Electroencephalogram Frequency-Tagging," *Philosophical Transactions of the Royal Society B: Biological Sciences* 369, no. 1658 (December 19, 2014): 20130393, doi.org/10.1098/rstb.2013.0393.

142 **employment center in Hässleholm, Sweden, granted disability benefits for a heavy metal "handicap"**: "Man Gets Sick Benefits for Heavy Metal Addiction," *The Local Sweden* (blog), June 19, 2007, accessed January 15, 2022, thelocal.se/20070619/7650/.

"I'm still very addicted": David Landes, "Swedish Heavy Metal Man: 'I'm Still Addicted,'" *The Local Sweden* (blog), January 11, 2013, thelocal.se/20130111/45566/.

music triggers the release of dopamine: Salimpoor et al., "Anatomically Distinct Dopamine Release During Anticipation and Experience of Peak Emotion to Music."

music and substance abuse: Genevieve A. Dingle et al., "The Influence of Music on Emotions and Cravings in Clients in Addiction Treatment: A Study of Two Clinical Samples," *The Arts in Psychotherapy* 45 (September 2015): 18–25, doi.org/10.1016/j.aip.2015.05.005.

"even for someone who is sort of obsessed with it . . . ": Zatorre, phone interview with author.

143 **some clients with schizophrenia, depression, bipolar disorder, or PTSD use music as a way of "sublimating" unwanted emotions or impulses**: Kevin Kirkland, Zoom interview with author, August 31, 2020.

"healthy excessive enthusiasms add to life, whereas addiction takes away from it": Mark Griffiths, "A 'Components' Model of Addiction Within a Biopsychosocial Framework," *Journal of Substance Use* 10, no. 4 (January 2005): 195, doi.org/10.1080/14659890500114359.

for more on musomania (music obsession) versus addiction, see also: Mark D. Griffiths, "Going for a Song: Can Listening to Music Be Addictive?," *Psychology Today*, May 22, 2014, psychologytoday.com/ca/blog/in-excess/201405/going-song.

"maladaptive music listening": Nicolas Schmuziger et al., "Is There Addiction to Loud Music? Findings in a Group of Non-Professional Pop/Rock Musicians," *Audiology Research* 2, no. 1 (July 3, 2012): e11, doi.org/10.4081/audiores.2012.e11.

Mark Griffiths, "Addictive Sounds?," *Psychology Today*, accessed January 15, 2022, psychologytoday.com/ca/blog/in-excess/202001/addictive-sounds.

144 **"at risk for dependence"**: Christine Ahrends, "Does Excessive Music Practicing Have Addiction Potential?," *Psychomusicology: Music, Mind, and Brain* 27, no. 3 (2017): 191–202, doi.org/10.1037/pmu0000188.

145 **"Our musical taste is greatly influenced by what we hear between the ages of twelve and twenty. . . "**: "Neuroscientist Explores 'Your Brain on Music,'" WUWM 89.7 FM–Milwaukee's NPR, December 26, 2018, wuwm.com/podcast/lake-effect-segments/2018-12-26/neuroscientist-explores-your-brain-on-music.

87 percent of American music consumers said they seldom strayed from the music they normally listened to: "Nielsen Music/MRC Data Midyear Report: U.S. 2020," July 2, 2020, 7.

"It's something you can make an effort to do, go out and hear new concerts . . . ": "Neuroscientist Explores 'Your Brain on Music.'"

"very similar to a post-traumatic stress disorder": Bittman, phone interview with author.

146 **post-traumatic stress disorder, symptoms**: "What Is PTSD?," American Psychiatric Association, accessed January 15, 2022, psychiatry.org/patients-families/ptsd/what-is-ptsd.

"Post-Traumatic Stress Disorder (PTSD): Symptoms and Causes," Mayo Clinic, accessed January 15, 2022, mayoclinic.org/diseases-conditions/post-traumatic-stress-disorder/symptoms-causes/syc-20355967.

"The Juilliard Effect": Daniel J. Wakin, "The Juilliard Effect: Ten Years Later," *The New York Times*, December 12, 2004, nytimes.com/2004/12/12/arts/music/the-juilliard-effect-ten-years-later.html.

"Classical Music's Alcohol Problem": Hugh Morris, "Confronting Classical Music's Alcohol Problem," I Care If You Listen, December 1, 2021, icareifyoulisten.com/2021/12/confronting-classical-music-alcohol-problem-casting-light-7/.

147 **"in such a way that I felt in harmony with the universe"**: Sam Roberts, "Mitchell L. Gaynor, 59, Dies; Oncologist and Author on Alternative Treatments," *The New York Times*, September 18, 2015, sec. Health, nytimes.com/2015/09/20/health/mitchell-l-gaynor-59-manhattan-oncologist-and-advocate-for-alternative-treatments-dies.html.

"a kind of cosmic symphony": Brian Greene, "Brian Greene, TED2012: 'Is Our Universe the Only Universe?,'" TED.com, February 2012, ted.com/talks/brian_greene_is_our_universe_the_only_universe/transcript.

148 **"I believe that sound can play a role in virtually any medical disorder..."**: Mitchell L. Gaynor, *The Healing Power of Sound: Recovery From Life-Threatening Illness Using Sound, Voice, and Music* (Boston: Shambhala, 2002).

Mitchell Gaynor suicide: Roberts, "Mitchell L. Gaynor, 59, Dies; Oncologist and Author on Alternative Treatments."

singing bowls are not an ancient Tibetan healing tool: Ben Joffe, "Tripping on Good Vibrations: Cultural Commodification and Tibetan Singing Bowls," Savage Minds: Notes and Queries on Anthropology, October 21, 2015, savageminds.org/2015/10/31/tripping-on-good-vibrations-cultural-commodification-and-tibetan-singing-bowls/.

Mitch Nur, "Singing Bowls: Separating Truth From Myth," Sound Travels, 2016, soundtravels.co.uk/a-Singing_Bowls__Separating_Truth_from_Myth-732.aspx.

"kindly stop mythologizing and exoticizing Tibetans..." Tenzin Dheden, "'Tibetan Singing Bowls' Are Not Tibetan. Sincerely, a Tibetan Person," *The Toronto Star*, February 18, 2020, sec. Contributors, thestar.com/opinion/contributors/2020/02/18/tibetan-singing-bowls-are-not-tibetan-sincerely-a-tibetan-person.html.

sound-bathing trend: Amitha Kalaichandran, "How Sound Baths Ended Up Everywhere," *The New York Times*, August 3, 2019, sec. Style, nytimes.com/2019/08/03/style/self-care/sound-baths.html.

149 **"People are looking for some sanctuary..."**: Michelle Robertson, "I Tried the New SF Meditation Trend That's Drawing People by the Thousands. Here's What It's Like," SFGate.com, October 28, 2017, sfgate.com/living/article/I-tried-the-insanely-popular-SF-sound-meditation-12303422.php.

cortisol and stress response: "How to Reduce Cortisol and Turn Down the Dial on Stress," Cleveland Clinic, August 27, 2020, health.clevelandclinic.org/how-to-reduce-cortisol-and-turn-down-the-dial-on-stress/.

"massage on the cellular level": "Soundshala Class Offerings," Soundshala, accessed January 15, 2022, soundshala.com/programs.

new-age remedies can do real harm: Skyler Bryce Johnson et al., "Use of Alternative Medicine for Cancer and Its Impact on Survival," *Journal of Clinical Oncology* 35, no. 15_suppl (May 20, 2017): e18175, doi.org/10.1200/JCO.2017.35.15_suppl.e18175.

"Cancer cells cannot maintain their structure...": Fabien Maman, *The Role of Music in the Twenty-First Century* (Redondo Beach, CA: Tama-Do Press, 1997).

beams of ultrasound can zap cancer cells: Robin McKie, "High-Power Sound Waves Used to Blast Cancer Cells," *The Observer*, October 31, 2015, sec. Science, theguardian.com/science/2015/oct/31/ultrasound-cancer-research-hifu-bone-trial.

8: All Together Now

152 **Babatunde Olatunji, the Nigerian master widely credited for turning North Americans on to hand drumming:** "Michael Babatunde Olatunji Biography," Encyclopedia of World Biography, notablebiographies.com/supp/Supplement-Mi-So/Olatunji-Michael-Babatunde.html.

154 **Christopher Small on "musicking":** Small, *Musicking*.

155 **"I'd like to buy the world a Coke":** "Creating 'I'd Like to Buy the World a Coke,'" The Coca-Cola Company, accessed January 13, 2022, coca-colacompany.com/company/history/creating-id-like-to-buy-the-world-a-coke.

ice-age flutes, bonding, and early survival: Helen Metella and Connie Edwards, "I Got Rhythm: The Science of Song," *The Nature of Things*, 2019, cbc.ca/natureofthings/features/humans-having-been-making-music-for-over-40000-years.

156 **"Music is in every aspect of life in our local communities...":** KellyAnne McGuire, "Strumming the Heartstrings," Chronogram Magazine, accessed January 14, 2022, chronogram.com/hudsonvalley/strumming-the-heartstrings/Content?oid=2169361.

Sascha Paladino, *Throw Down Your Heart*, documentary film, 2008, youtube.com/watch?v=sJt6jnoxT8A.

the akonting, ancestor of the banjo: Scott V. Linford, "Historical Narratives of the Akonting and Banjo," *Ethnomusicology Review* 2, no. 22 (July 27, 2014), ethnomusicologyreview.ucla.edu/content/akonting-history.

the akonting in Gambian lore: Paladino, *Throw Down Your Heart*.

"Bela Fleck and Toumani Diabate: Banjo Roots," *NPR*, April 13, 2009, sec. Studio Sessions, npr.org/templates/story/story.php?storyId=103002655.

157 **Jeremy Dutcher and Maggie Paul:** "About: Jeremy Dutcher," Jeremy Dutcher, accessed January 13, 2022, jeremydutcher.com/about/.

Ryan McNutt, "A Dialogue Spanning Time and Tradition," Dalhousie News, September 18, 2018, dal.ca/news/2018/05/10/jeremy-dutcher-brings-historic-indigenous-recordings-to-life.html.

"We brought the music back, we brought the drum back...": Jeremy Dutcher, "Eqpahak," track from *Wolastoqiyik Lintuwakonawa* (Universal Music, 2019).

Mothers of every culture sing to their babies: Sandra E. Trehub, Anna M. Unyk, and Laurel J. Trainor, "Maternal Singing in Cross-Cultural Perspective," *Infant Behavior and Development* 16, no. 3 (July 1993): 285–95, doi.org/10.1016/0163-6383(93)80036-8.

158 **lullabies soothe mothers as well as their babies:** Dennie Palmer Wolf, "Lullaby: Being Together, Being Well," WolfBrown, May 2017, wolfbrown.com/post/lullaby-being-together-being-well.

Laura K. Cirelli, Zuzanna B. Jurewicz, and Sandra E. Trehub, "Effects of Maternal Singing Style on Mother–Infant Arousal and Behavior," *Journal of Cognitive Neuroscience* 32, no. 7 (July 1, 2020): 1213–20, doi.org/10.1162/jocn_a_01402.

Shmuel Arnon et al., "Maternal Singing During Kangaroo Care Led to Autonomic Stability in Preterm Infants and Reduced Maternal Anxiety," *Acta Paediatrica* 103, no. 10 (October 2014): 1039–44, doi.org/10.1111/apa.12744.

hearing development in fetuses and newborns: Peter G. Hepper and B. Sara Shahidullah, "Development of Fetal Hearing," *Archives of Disease in Childhood. Fetal and Neonatal Edition* 71, no. 2 (September 1994): F81–87, NCBI.nlm.nih .gov/pmc/articles/PMC1061088/.

Stanley N. Graven and Joy V. Browne, "Auditory Development in the Fetus and Infant," *Newborn and Infant Nursing Reviews* 8, no. 4 (December 1, 2008): 187–93, doi.org/10.1053/j.nainr.2008.10.010.

Children's Hospital of Philadelphia, "The Senses of a Newborn," Children's Hospital of Philadelphia (The Children's Hospital of Philadelphia, August 23, 2014), chop.edu/conditions-diseases/newborn-senses.

Newborns can remember songs they heard from the watery haven of the womb: Carolyn Granier-Deferre et al., "A Melodic Contour Repeatedly Experienced by Human Near-Term Fetuses Elicits a Profound Cardiac Reaction One Month After Birth," ed. Georges Chapouthier, *PLOS ONE* 6, no. 2 (February 23, 2011): e17304, doi.org/10.1371/journal.pone.0017304.

159 **infants remained calm for twice as long when they heard singing versus speech:** Mariève Corbeil, Sandra E. Trehub, and Isabelle Peretz, "Singing Delays the Onset of Infant Distress," *Infancy* 21, no. 3 (May 2016): 373–91, doi.org/10.1111/infa.12114.

distressed infants smiled more and showed calmer responses: Laura K. Cirelli and Sandra E. Trehub, "Familiar Songs Reduce Infant Distress," *Developmental Psychology* 56, no. 5 (May 2020): 861–68, doi.org/10.1037/dev0000917.

overview of oxytocin: Beth Azar, "Oxytocin's Other Side," American Psychological Association, March 2011, apa.org/monitor/2011/03/oxytocin.

Thomas R. Insel et al., "Oxytocin, Vasopressin, and the Neuroendocrine Basis of Pair Bond Formation," in *Vasopressin and Oxytocin*, ed. Hans H. Zingg, Charles W. Bourque, and Daniel G. Bichet, vol. 449, *Advances in Experimental Medicine and Biology* (Boston, MA: Springer US, 1998), 215–24, doi.org/10.1007 /978-1-4615-4871-3_28.

"You'll be happier...": Paul Zak, *Trust, Morality—and Oxytocin?*, TED.com, 2011, ted.com/talks/paul_zak_trust_morality_and_oxytocin.

160 **"experience more empathy," "be your true, uninhibited self":** "Connekt," VeroLabs.com, accessed January 13, 2022, Verolabs.com/product/connekt/1.

listening to music increased oxytocin in cardiac patients: Ulrica Nilsson, "Soothing Music Can Increase Oxytocin Levels During Bed Rest After Open-Heart Surgery: A Randomised Control Trial," *Journal of Clinical Nursing* 18, no. 15 (August 2009): 2153–61, doi.org/10.1111/j.1365-2702.2008.02718.x.

choir members had higher oxytocin levels after half an hour of singing: Gunter Kreutz, "Does Singing Facilitate Social Bonding?," *Music and Medicine* 6, no. 2 (October 25, 2014): 51, doi.org/10.47513/mmd.v6i2.180.

"People got carried away with the idea of the cuddle hormone": Azar, "Oxytocin's Other Side."

oxytocin and envy, schadenfreude, aggression to other groups, sensitivity to social cues: Simone G. Shamay-Tsoory et al., "Intranasal Administration of Oxytocin Increases Envy and Schadenfreude (Gloating)," *Biological Psychiatry* 66, no. 9 (November 2009): 864–70, doi.org/10.1016/j.biopsych.2009.06.009.

C. K. W. De Dreu et al., "Oxytocin Promotes Human Ethnocentrism," *Proceedings of the National Academy of Sciences* 108, no. 4 (January 25, 2011): 1262–66, doi.org/10.1073/pnas.1015316108.

Azar, "Oxytocin's Other Side."

oxytocin can spike in response to stress: Yuki Takayanagi and Tatsushi Onaka, "Roles of Oxytocin in Stress Responses, Allostasis and Resilience," *International Journal of Molecular Sciences* 23, no. 1 (December 23, 2021): 150, doi.org/10.3390/ijms23010150.

161 **"it turns out that you trust them more—you are more likely to befriend them . . .":** McMaster LIVELab, *About the LIVELab* (Hamilton, Ontario, 2018), youtube.com/watch?v=odccx59Do8o.

"muscular bonding": William Hardy McNeill, *Keeping Together in Time: Dance and Drill in Human History* (Cambridge, MA: Harvard University Press, 1995), 2.

study after study has linked synchronous movements to stronger social ties:

Reneeta Mogan, Ronald Fischer, and Joseph A. Bulbulia, "To Be in Synchrony or Not? A Meta-Analysis of Synchrony's Effects on Behavior, Perception, Cognition and Affect," *Journal of Experimental Social Psychology* 72 (September 2017): 13–20, doi.org/10.1016/j.jesp.2017.03.009.

Matthew H. Woolhouse, Dan Tidhar, and Ian Cross, "Effects on Inter-Personal Memory of Dancing in Time With Others," *Frontiers in Psychology* 7 (February 23, 2016): 167, doi.org/10.3389/fpsyg.2016.00167.

162 **"weaken the psychological boundaries between the self and the group":** Scott S. Wiltermuth and Chip Heath, "Synchrony and Cooperation," *Psychological Science* 20, no. 1 (January 2009): 1–5, doi.org/10.1111/j.1467-9280.2008.02253.x.

music in particular brings us closer when we move as one: Jan Stupacher et al., "Music Strengthens Prosocial Effects of Interpersonal Synchronization—If You Move in Time With the Beat," *Journal of Experimental Social Psychology* 72 (September 1, 2017): 39–44, doi.org/10.1016/j.jesp.2017.04.007.

drumming, group connection, heartbeat synchrony: Ilanit Gordon et al., "Physiological and Behavioral Synchrony Predict Group Cohesion and Performance," *Scientific Reports* 10, no. 1 (December 2020): 8484, doi.org/10.1038/s41598-020-65670-1.

fourteen-month-old infants, rhythm, and bonding: Laura K. Cirelli, Stephanie J. Wan, and Laurel J. Trainor, "Fourteen-Month-Old Infants Use Interpersonal Synchrony as a Cue to Direct Helpfulness," *Philosophical Transactions of the Royal Society B: Biological Sciences* 369, no. 1658 (December 19, 2014): 20130400, doi.org/10.1098/rstb.2013.0400.

Laura K. Cirelli, Kathleen M. Einarson, and Laurel J. Trainor, "Interpersonal Synchrony Increases Prosocial Behavior in Infants," *Developmental Science* 17, no. 6 (November 2014): 1003–11, doi.org/10.1111/desc.12193.

spectators' brainwaves begin to match those of the performer: Yingying Hou et al., "The Averaged Inter-Brain Coherence Between the Audience and a Violinist Predicts the Popularity of Violin Performance," *NeuroImage* 211 (May 2020): 116655, doi.org/10.1016/j.neuroimage.2020.116655.

163 **brain synchrony among guitarists performing together:** Ulman Lindenberger et al., "Brains Swinging in Concert: Cortical Phase Synchronization While Playing Guitar," *BMC Neuroscience* 10, no. 1 (December 2009): 22, doi.org/10.1186/1471-2202-10-22.

Johanna Sänger, Viktor Müller, and Ulman Lindenberger, "Intra- and Interbrain Synchronization and Network Properties When Playing Guitar in Duets," *Frontiers in Human Neuroscience* 6 (2012), doi.org/10.3389/fnhum.2012.00312.

"two brains make one synchronized mind": Naoyuki Osaka et al., "How Two Brains Make One Synchronized Mind in the Inferior Frontal Cortex: FNIRS-Based Hyperscanning During Cooperative Singing," *Frontiers in Psychology* 6 (2015), frontiersin.org/article/10.3389/fpsyg.2015.01811.

LIVELab at McMaster University: Léo Charbonneau, "McMaster U's Interactive Research Lab Is Much More Than a Concert Hall," *University Affairs* (blog), February 11, 2015, universityaffairs.ca/news/news-article/mcmaster-us-interactive-research-lab-much-concert-hall/.

brainwave synchrony in audience members: Adela Talbot, "Live Crowds Find Their Groove Together," University of Western Ontario, April 9, 2018, ssc.uwo.ca/news/2018/Grahn_live_music_groove.html.

Grahn, email correspondence with author.

Rhythms and songs speed up the bonding process, dialing us in to a central hub of connection: Talbot, "Live Crowds Find Their Groove Together."

Daniel Weinstein et al., "Singing and Social Bonding: Changes in Connectivity and Pain Threshold as a Function of Group Size," *Evolution and Human Behavior* 37, no. 2 (March 2016): 152–58, doi.org/10.1016/j.evolhumbehav.2015.10.002.

164 **"There's this unifying force that comes from the music":** Elizabeth Landau, "This Is Your Brain on Music," CNN, January 23, 2018, cnn.com/2013/04/15/health/brain-music-research/index.html.

Dalai Lama event in Vancouver: Doug Ward, "Humble Dalai Lama Simply a Superstar," *Vancouver Sun*, September 11, 2006.

"Dalai Lama Wows Thousands," DalaiLama.com (The 14th Dalai Lama, September 11, 2006), dalailama.com/news/2006/dalai-lama-wows-thousands/amp.

165 **singing with others may be fundamental to our biology:** Sarah Keating, "The World's Most Accessible Stress Reliever," BBC, May 18, 2020, bbc.com/future/article/20200518-why-singing-can-make-you-feel-better-in-lockdown.

choir singing and social connection: Weinstein et al., "Singing and Social Bonding."

music, children, and "empathy education": Tal-Chen Rabinowitch, Ian Cross, and Pamela Burnard, "Long-Term Musical Group Interaction Has a Positive Influence on Empathy in Children," *Psychology of Music* 41, no. 4 (July 2013): 484–98, doi.org/10.1177/0305735612440609.

166 **"Through music, streamed live...":** Andrea Bocelli, *Andrea Bocelli:Music for Hope—Live From Duomo di Milano*, 2020, youtube.com/watch?v=huTUoek4LgU.

coronavirus, lockdown, music, and connection: Alasdair Sandford, "Coronavirus: Half of Humanity on Lockdown in 90 Countries," Euronews, April 2, 2020, euronews.com/2020/04/02/coronavirus-in-europe-spains-death-toll-hits-10-000-after-record-950-new-deaths-in-24-hou.

Matt Clinch, "Italians Are Singing Songs From Their Windows to Boost Morale During Coronavirus Lockdown," CNBC, March 14, 2020, cnbc.com/2020/03/14/coronavirus-lockdown-italians-are-singing-songs-from-balconies.html.

Andy Greene, "See Neil Young's Fourth 'Fireside Sessions' Home Concert," *Rolling Stone*, May 7, 2020, rollingstone.com/music/music-news/see-neil-youngs-fourth-fireside-sessions-concert-995377/.

Gary Dinges and Patrick Ryan, "Coronavirus Concerts: Radiohead, Alicia Keys, More Sharing Free Shows," *USA Today*, March 24, 2020, usatoday.com/story/entertainment/music/2020/03/24/coronavirus-concerts-rob-thomas-garth-brooks-john-legend-diplo-and-more-perform-live-home/2907314001/.

Kathryn Shattuck, "Yo-Yo Ma Tries to Bring Us Comfort and Hope," *The New York Times*, June 9, 2020, sec. Arts, nytimes.com/2020/06/09/arts/music/yo-yo-ma-favorite-things.html.

167 **"We go back to these roots of [the] innate things that bring us connection and strength":** Keating, "The World's Most Accessible Stress Reliever."

Aleppo, continuously inhabited for at least five thousand years: The Editors of Encyclopaedia Britannica, "Aleppo: History, Map, Citadel, & Facts," Britannica.com, accessed January 13, 2022, britannica.com/place/Aleppo.

Aleppo and its warmhearted people were devastated by civil war: Starting in 2012, warfare forced 6.7 million Syrians to flee as refugees, obliterated 10 percent of Aleppo's historic buildings, and left more than half with moderate to severe damage:

"By the Numbers: Syrian Refugees Around the World," FRONTLINE, PBS, accessed February 3, 2022, pbs.org/wgbh/frontline/article/numbers-syrian-refugees-around-world/.

UNESCO World Heritage Centre, "Five Years of Conflict—the State of Cultural Heritage in the Ancient City of Aleppo," UNESCO World Heritage Centre, accessed February 3, 2022, whc.unesco.org/en/activities/946/.

168 **Aleppo citadel and throne room:** Abdallah Hadjar, *Historical Monuments of Aleppo,* translated by Khaled Al-Jbaili (Aleppo: Automobile and Touring Club of Syria, 2000), 14.

"a metaphor for human connectedness, creativity, and imagination": Theodore Levin, phone interview with author, September 22, 2002.

"axis of evil": History.com editors, "George W. Bush Describes Iraq, Iran and North Korea as 'Axis of Evil,'" History.com, accessed January 13, 2022, history.com/this-day-in-history/bush-describes-iraq-iran-north-korea-as-axis-of-evil.

169 **"Blue as the Turquoise Night of Neyshabur":** "Yo-Yo Ma Performs on the Silk Road for Aga Khan Award," Aga Khan Development Network, November 6, 2001, akdn.org/press-release/yo-yo-ma-performs-silk-road-aga-khan-award.

"I've always started from the premise . . . ": Yo-Yo Ma, phone interview with author, September 23, 2002.

170 **David Byrne calls top-forty the "fast food" of music:** David Byrne, "Music: Crossing Music's Borders in Search of Identity; 'I Hate World Music,'" *The New York Times*, October 3, 1999, sec. Books, nytimes.com/1999/10/03/arts/music-crossing-musics-borders-in-search-of-identity-i-hate-world-music.html.

171 **children who shared music with their parents feel closer to them later on:** Sandi D. Wallace and Jake Harwood, "Associations Between Shared Musical Engagement and Parent–Child Relational Quality: The Mediating Roles of Interpersonal Coordination and Empathy," *Journal of Family Communication* 18, no. 3 (July 3, 2018): 202–16, doi.org/10.1080/15267431.2018.1466783.

173 **"The energy of the crowd kind of feeds us...":** Bloco Energia, *Bloco Energia: Davie Street Pride Party*, 2016, youtube.com/watch?v=uflH-DCQjTY.

9: The Beat Goes On

175 **strumming Radiohead songs on the ukulele:** Rhodri Marsden, "Am I the Only Person in the World Who Hates the Ukulele?," *The Independent*, June 21, 2014, independent.co.uk/arts-entertainment/music/features/rhodri-marsden-am-i-the-only-person-in-the-world-who-hates-the-ukulele-9553063.html.

Jumpin' Jim's Ukulele Island: Jumpin' Jim's Ukulele Island (Los Angeles, CA; Milwaukee, WI: Flea Market Music; Distributed by Hal Leonard, 2004).

177 **gunpowder instructions in *Every Boy's Book*:** Edmund Routledge, *Every Boy's Book: A Complete Encyclopaedia of Sports and Amusements* (G. Routledge and Sons, 1869), 693.

178 **bilingualism and delayed dementia:** Víctor Costumero et al., "A Cross-Sectional and Longitudinal Study on the Protective Effect of Bilingualism Against Dementia Using Brain Atrophy and Cognitive Measures," *Alzheimer's Research & Therapy* 12, no. 1 (December 2020): 11, doi.org/10.1186/s13195-020-0581-1.

Daniela Perani et al., "The Impact of Bilingualism on Brain Reserve and Metabolic Connectivity in Alzheimer's Dementia," *Proceedings of the National Academy of Sciences* 114, no. 7 (February 14, 2017): 1690–95, doi.org/10.1073/pnas.1610909114.

music training tied to better cognition later in life: Brenda Hanna-Pladdy and Alicia MacKay, "The Relation Between Instrumental Musical Activity and Cognitive Aging," *Neuropsychology* 25, no. 3 (2011): 378–86, doi.org/10.1037/a0021895.

Brenda Hanna-Pladdy and Byron Gajewski, "Recent and Past Musical Activity Predicts Cognitive Aging Variability: Direct Comparison With General Lifestyle Activities," *Frontiers in Human Neuroscience* 6 (2012), doi.org/10.3389/fnhum.2012.00198.

T. White-Schwoch et al., "Older Adults Benefit From Music Training Early in Life: Biological Evidence for Long-Term Training-Driven Plasticity," *Journal of Neuroscience* 33, no. 45 (November 6, 2013): 17667–74, doi.org/10.1523/JNEUROSCI.2560-13.2013.

179 **overview of dementia:** "Dementia: Key Facts," World Health Organization, accessed January 12, 2022, who.int/news-room/fact-sheets/detail/dementia.

music as a "promising cognitive intervention" for older adults: Ryan Sutcliffe, Kangning Du, and Ted Ruffman, "Music Making and Neuropsychological Aging: A Review," *Neuroscience & Biobehavioral Reviews* 113 (June 2020): 479–91, doi.org/10.1016/j.neubiorev.2020.03.026.

180 **twin study suggests "playing an instrument has a positive influence on neuroplasticity, regardless of what age one begins":** M. Alison Balbag, Nancy L. Pedersen, and Margaret Gatz, "Playing a Musical Instrument as a Protective

Factor Against Dementia and Cognitive Impairment: A Population-Based Twin Study," *International Journal of Alzheimer's Disease* 2014 (2014): 1–6, doi.org/10.1155/2014/836748.

latecomers to music put to the test: J. A. Bugos et al., "Individualized Piano Instruction Enhances Executive Functioning and Working Memory in Older Adults," *Aging & Mental Health* 11, no. 4 (July 2007): 464–71, doi.org/10.1080/13607860601086504.

"If you're just thinking the same thoughts . . .": CopperCat Band, *Daniel Levitin Interview—From Boomer Bands Documentary.*

181 **compared to non-musicians, instrumentalists had younger-looking brains:** Lars Rogenmoser et al., "Keeping Brains Young With Making Music," *Brain Structure and Function* 223, no. 1 (January 2018): 297–305, doi.org/10.1007/s00429-017-1491-2.

meaning of "amateur": "Amateur: Meaning & Definition for UK English," Lexico Dictionaries, English, accessed January 12, 2022, lexico.com/definition/amateur.

"Amateur: Etymology, Origin and Meaning," Etymonline: Online Etymology Dictionary, accessed January 12, 2022, etymonline.com/word/amateur.

Gen-Xers want to learn a musical instrument before they die: Lauren Slocum, "101+ Bucket List Ideas of Things to Do Before You Die," Ranker.com, accessed January 12, 2022, ranker.com/list/bucket-list-ideas/lauren-slocum.

"Choose the right instrument . . .": CopperCat Band, *Daniel Levitin Interview—From Boomer Bands Documentary.*

182 **there's no age limit for learning to play the cello, viola, or violin:** Miranda Wilson, "It's Never Too Late to Pick Up an Instrument—Here Are 5 Tips for Adult Beginners," *Strings Magazine* (blog), December 26, 2018, stringsmagazine.com/its-never-too-late-to-pick-up-an-instrument-here-are-5-tips-for-adult-beginners/.

183 **"There's never been a better time to learn guitar . . .":** Peter Kennedy, phone interview with author, December 17, 2020.

184 **loneliness declared a public health crisis, equivalent to smoking:** Vivek Murthy, "Loneliness Was a Public Health Crisis Long Before Social Distancing. Here's How We Can Solve It," *Time*, March 26, 2020, time.com/collection/apart-not-alone/5809171/loneliness-epidemic-united-states/.

Julianne Holt-Lunstad, Timothy B. Smith, and J. Bradley Layton, "Social Relationships and Mortality Risk: A Meta-Analytic Review," *PLOS Medicine* 7, no. 7 (July 27, 2010): e1000316, doi.org/10.1371/journal.pmed.1000316.

"Why Loneliness Can Be as Unhealthy as Smoking 15 Cigarettes a Day," CBC, August 16, 2017, cbc.ca/news/health/loneliness-public-health-psychologist-1.4249637.

among American retirees, 43 percent reported loneliness: Carla M. Perissinotto, Irena Stijacic Cenzer, and Kenneth E. Covinsky, "Loneliness in Older Persons: A Predictor of Functional Decline and Death," *Archives of Internal Medicine* 172, no. 14 (July 23, 2012), doi.org/10.1001/archinternmed.2012.1993.

loneliness leaves its mark on the brain: R. Nathan Spreng et al., "The Default Network of the Human Brain Is Associated With Perceived Social Isolation," *Nature Communications* 11, no. 1(December 15, 2020): 6393, doi.org/10.1038/s41467-020-20039-w.

"Scientists Show What Loneliness Looks Like in the Brain," The Neuro, December 15, 2020, mcgill.ca/neuro/channels/news/scientists-show-what-loneliness-looks-brain-325504.

The researchers noted that either loneliness causes changes to the brain's default network or—chicken or egg—it could be that having more density in this brain region makes people more vulnerable to loneliness.

Chronic loneliness increases the risk of dementia by 50 percent: National Academies of Sciences, Engineering, and Medicine, *Social Isolation and Loneliness in Older Adults: Opportunities for the Health Care System* (Washington, DC: The National Academies Press, 2020).

185 **Community of Voices choir linked to improvements in loneliness:** Julene K. Johnson et al., "A Community Choir Intervention to Promote Well-Being Among Diverse Older Adults: Results From the Community of Voices Trial," *The Journals of Gerontology Series B: Psychological Sciences and Social Sciences* 75, no. 3 (February 2020): 549–59, doi.org/10.1093/geronb/gby132.

"Comprehensive Program Manual," Community of Voices, accessed January 12, 2022, cov.ucsf.edu/comprehensive-program-manual.

"singing broke the ice better": "Singing's Secret Power: The Ice-Breaker Effect," University of Oxford, October 28, 2015, ox.ac.uk/news/2015-10-28-singing's-secret-power-ice-breaker-effect-1.

Eiluned Pearce, Jacques Launay, and Robin I. M. Dunbar, "The Ice-Breaker Effect: Singing Mediates Fast Social Bonding," *Royal Society Open Science* 2, no. 10: 150221, accessed January 12, 2022, doi.org/10.1098/rsos.150221.

Family attitudes about music, combined with musical activities at home, predict with 74-percent accuracy which children will choose to make music: Demorest, Kelley, and Pfordresher, "Singing Ability, Musical Self-Concept, and Future Music Participation."

186 **William A. Mathieu: even severely pitch-challenged people can learn to sing:** Megan Lane, "Can the Tone Deaf Learn to Sing?," *BBC News*, January 10, 2011, sec. Magazine, bbc.com/news/magazine-12127843.

"can't sing" choirs and "Tone Deaf? No Way!": "The Choir Who Can't Sing (George Bevan)," Musical U, February 14, 2017, musical-u.com/learn/guest-post-the-choir-who-cant-sing-george-bevan/.

"Choir 'Sings Like No-One's Listening,'" *BBC News*, January 29, 2019, bbc.com/news/av/uk-england-oxfordshire-47030582.

"Tone Deaf? No Way!," Morley College, accessed February 3, 2022, findcourses.co.uk/training/morley-college/tone-deaf-no-way-1169881.

group singing and depression, Parkinson's, stroke: Bill Ahessy, "The Use of a Music Therapy Choir to Reduce Depression and Improve Quality of Life in Older Adults—a Randomized Control Trial," *Music and Medicine* 8, no. 1 (January 31, 2016): 17–28, doi.org/10.47513/mmd.v8i1.451.

Laura Fogg-Rogers et al., "Choral Singing Therapy Following Stroke or Parkinson's Disease: An Exploration of Participants' Experiences," *Disability and Rehabilitation* 38, no. 10 (May 7, 2016): 952–62, doi.org/10.3109/09638288.2015.1068875.

prescribing singing for trauma: Galya Chatterton, personal communication with author, January 13, 2022.

187 **"Somewhere along the way, many of us became musical spectators instead of participants":** "Rhythm 'n' Roots: About," Rhythm and Roots, accessed January 12, 2022, rhythmandroots.ca/about/.

188 **Barbra Streisand, Back to Brooklyn concert, Ottawa:** Denis Armstrong, "Barbra Streisand Royally Good at Scotiabank," *Ottawa Sun*, accessed January 25, 2022, ottawasun.com/2012/10/18/barbara-streisand-at-scotiabank.

botticelli101, *Chris Botti Live in Ottawa at Scotiabank Place—Barbra Streisand Back to Brooklyn Tour 2012*, 2012, youtube.com/watch?v=oc1wovojwhi.

189 **"every single person in our choir sings from their heart...":** Madeleine Pouliot, phone interview with author, January 29, 2021.

"It's what keeps me sane": Bruce Deachman, "The Chorister," *Ottawa Citizen*, March 6, 2011.

190 **Parkinson's epidemic, risk factors:** E. Ray Dorsey et al., "The Emerging Evidence of the Parkinson Pandemic," *Journal of Parkinson's Disease* 8, no. s1 (2018): s3–8, doi.org/10.3233/jpd-181474.

"Parkinson's Disease Risk Factors and Causes," Johns Hopkins Medicine, accessed January 12, 2022, hopkinsmedicine.org/health/conditions-and-diseases/parkinsons-disease/parkinsons-disease-risk-factors-and-causes.

Parkinson's involves a drop in dopamine: Insha Zahoor, Amrina Shafi, and Ehtishamul Haq, "Pharmacological Treatment of Parkinson's Disease," in *Parkinson's Disease: Pathogenesis and Clinical Aspects*, ed. Thomas B. Stoker and Julia C. Greenland (Brisbane (AU): Codon Publications, 2018), ncbi.nlm.nih.gov/books/NBK536726/.

"Music has something that revs up our motor function: rhythm": Grahn, email correspondence with author.

"What the music does is to trigger the movements trapped inside": La Trobe University, *Dancing With Parkinson's*, 2014, youtube.com/watch?v=sGt1ci9E3gs.

neurologists are studying whether playing a musical instrument can improve fine-motor skills in Parkinson's: Greg Glasgow, "A Rhythmic Approach to Music Therapy for Parkinson's Patients," October 22, 2021, news.cuanschutz.edu/medicinermusic-therapy-for-parkinsons-patients.

191 **in people with mild to moderate Parkinson's, unmistakable improvements in those who danced:** Karolina A. Bearss and Joseph F. X. DeSouza, "Parkinson's Disease Motor Symptom Progression Slowed With Multisensory Dance Learning Over 3-Years: A Preliminary Longitudinal Investigation," *Brain Sciences* 11, no. 7 (July 2021): 895, doi.org/10.3390/brainsci11070895.

normal progression of Parkinson's "is not seen in our dancers with Parkinson's...": Christine Sismondo, "Study Says Dancing Helps Slow Progression of Symptoms of Parkinson's Disease," *The Toronto Star*, July 26, 2021, sec. Opinion, thestar.com/life/health_wellness/opinion/2021/07/26/study-says-dancing-helps-slow-progression-of-symptoms-of-parkinsons-disease.html.

history of Dance for Parkinson's: "Dance for PD: About Us," Dance for PD, accessed January 12, 2022, danceforparkinsons.org/about-the-program.

Patricia Needle and Dance for Parkinson's: "Peace About Life: Dancing With Parkinson's," dNaga Dance Co., accessed January 12, 2022, dnaga.org/dances/peace-about-life.

"I was deliriously happy...": Patricia Needle, "Dance Redemption," KQED, April 25, 2011, kqed.org/perspectives/201104250735/dance-redemption.

192 **Marta González Saldaña and *Swan Lake*:** Agencia EFE, *Homenaje a una bailarina con alzheimer del "Lago de los Cisnes,"* 2020, youtube.com/watch?v=mz8hsM8DCqs.

the cognitive ability to enjoy music remains remarkably intact, even in people who are losing touch with their environment: J.B. King et al., "Increased Functional Connectivity After Listening to Favored Music in Adults With Alzheimer Dementia," *The Journal of Prevention of Alzheimer's Disease*, 2018, 1–7, doi.org/10.14283/jpad.2018.19.

"Music Activates Regions of the Brain Spared by Alzheimer's Disease," University of Utah, April 27, 2018, healthcare.utah.edu/publicaffairs/news/2018/04/alzheimer.php.

music that holds personal meaning may stimulate vital brain connections in people with mild cognitive impairment or early Alzheimer's: Corinne E. Fischer et al., "Long-Known Music Exposure Effects on Brain Imaging and Cognition in Early-Stage Cognitive Decline: A Pilot Study," *Journal of Alzheimer's Disease: JAD* 84, no. 2 (2021): 819–33, doi.org/10.3233/JAD-210610.

193 **"Typically, it's very difficult to show positive brain changes in Alzheimer's patients...":** "On Repeat: Listening to Favourite Music Improves Brain Plasticity, Cognitive Performance of Alzheimer's Patients: Toronto Study," University of Toronto—Faculty of Music, November 9, 2021, music.utoronto.ca/mob-we-live-here.php?exID=189.

short-term memory is usually the first to go: "When You Should Seek Help for Memory Loss," Mayo Clinic, accessed January 12, 2022, mayoclinic.org/diseases-conditions/alzheimers-disease/in-depth/memory-loss/art-20046326.

types of memory: Charles Stangor and Jennifer Walinga, "Memories as Types and Stages," October 17, 2014, opentextbc.ca/introductiontopsychology/chapter/8-1-memories-as-types-and-stages/.

music is "incredibly resistant to forgetting...": Adriana Barton, "What Happened When Victoria Teens Joined Seniors With Dementia in a Choir," *The Globe and Mail*, August 10, 2018, theglobeandmail.com/life/health-and-fitness/article-teens-and-seniors-with-dementia-joined-forces-in-a-choir-then/.

194 **"not even macaroni and cheese...":** Barton.

195 **playing music to connect with family members in later stages of dementia:** Jonathan Graff-Radford, "How Music Can Help People With Alzheimer's," Mayo Clinic, accessed January 12, 2022, mayoclinic.org/diseases-conditions/alzheimers-disease/expert-answers/music-and-alzheimers/faq-20058173.

benefits of Music and Memory program in California nursing homes: Debra Bakerjian et al., "The Impact of Music and Memory on Resident Level Outcomes in California Nursing Homes," *Journal of the American Medical Directors Association* 21, no. 8 (August 1, 2020): 1045-1050.e2, doi.org/10.1016/j.jamda.2020.01.103.

196 **"In all of us, there is a hunger, marrow deep...":** Alex Haley, "What *Roots* Means to Me," *Reader's Digest*, May 1977.

10: Fumbling Towards Ecstasy

198 **history of Polish Christmas carols and traditions:** Mieczyslaw Giergielewicz, *Introduction to Polish Versification* (Philadelphia: University of Pennsylvania Press, 1970), degruyter.com/doi/book/10.9783/9781512816259.

Kirsty Hardial, "All I Want for Christmas Is... Tunes," The First News, December 26, 2018, thefirstnews.com/article/all-i-want-for-christmas-istunes-tfn-takes-a-look-at-the-family-tradition-of-carol-singing-3950.

199 **history of Basilica of St. Francis of Assisi, Krakow, Poland:** "Basilica of St. Francis of Assisi—Kraków," Introducing Krakow, accessed January 11, 2022, introducingkrakow.com/basilica-st-francis-of-assisi.

use of ocher in early human burial: Erella Hovers et al., "An Early Case of Color Symbolism: Ochre Use by Modern Humans in Qafzeh Cave," *Current Anthropology* 44, no. 4 (August 2003): 491–522, doi.org/10.1086/375869.

religious and spiritual beliefs have been traced to a circuit in the brain's oldest and innermost region: Michael A. Ferguson et al., "A Neural Circuit for Spirituality and Religiosity Derived From Patients With Brain Lesions," *Biological Psychiatry*, June 2021, S0006322321014037, doi.org/10.1016/j.biopsych.2021.06.016.

"Researchers Identify Brain Circuit for Spirituality," EurekAlert!, July 1, 2021, eurekalert.org/news-releases/735597.

200 **"oneness," happiness, and life satisfaction:** Laura Marie Edinger-Schons, "Oneness Beliefs and Their Effect on Life Satisfaction," *Psychology of Religion and Spirituality* 12, no. 4 (November 2020): 428–39, doi.org/10.1037/rel0000259.

music in Tibetan Buddhism: Jeffrey W. Cupchik, "Buddhism as Performing Art: Visualizing Music in the Tibetan Sacred Ritual Music Liturgies," *Yale Journal of Music & Religion* 1, no. 1 (February 5, 2015), doi.org/10.17132/2377-231X.1010.

Hilary Herdman, "Tibetan Ritual Music," Samye Institute, July 30, 2019, samyeinstitute.org/sciences/arts/tibetan-ritual-music/.

201 **melody processed in the right auditory cortex; speech in the left:** Philippe Albouy et al., "Distinct Sensitivity to Spectrotemporal Modulation Supports Brain Asymmetry for Speech and Melody," *Science*, February 28, 2020, doi.org/10.1126/science.aaz3468.

No wonder hymns are easier to remember than scripture, and songs carry special meaning in cultures worldwide: Matthew D. Schulkind, "Is Memory for Music Special?," *Annals of the New York Academy of Sciences* 1169, no. 1 (July 2009): 216–24, doi.org/10.1111/j.1749-6632.2009.04546.x.

Gregory Crowther, "Using Science Songs to Enhance Learning: An Interdisciplinary Approach," ed. Marshall David Sundberg, *CBE—Life Sciences Education* 11, no. 1 (March 2012): 26–30, doi.org/10.1187/cbe.11-08-0068.

Sandra E. Trehub, Judith Becker, and Iain Morley, "Cross-Cultural Perspectives on Music and Musicality," *Philosophical Transactions of the Royal Society B: Biological Sciences* 370, no. 1664 (March 19, 2015): 20140096, doi.org/10.1098/rstb.2014.0096.

the medicine songs known as ikaros: Ikaros (or icaros) are just one of several kinds of medicine songs used by the Shipibo-Konibo people (also spelled Shipibo-Conibo), though outsiders tend to refer to all of these songs as ikaros. For more on Shipibo medicine songs and the myth of Shipibo embroideries as "singable designs," see:

Bernd Brabec de Mori, "The Magic of Song, the Invention of Tradition, and the Structuring of Time Among the Shipibo (Peruvian Amazon)," ed. Greta Lechleitner and Christian Liebl, *Jahrbuch des Phonogrammarchivs der Österreichischen Akademieder Wissenschaften* [Yearbook of the Phonogrammarchiv at the Austrian Academy of Sciences] 2 (January 2011): 169–92.

203 **cymatics, the study of visual patterns made by sound:** John Stuart Reid, "Cymatics: Sound Science of the Future," Rubin Museum of Art, accessed January 11, 2022, rubinmuseum.org/spiral/cymatics-sound-science-of-the-future.

ayahuasca and brain activity: Christopher Timmermann et al., "Neural Correlates of the DMT Experience Assessed With Multivariate EEG," *Scientific Reports* 9, no. 1 (November 19, 2019): 16324, doi.org/10.1038/s41598-019-51974-4.

"chaotic patterns" of brain activity: Ryan O'Hare, "Ayahuasca Compound Changes Brainwaves to Vivid 'Waking-Dream' State," Imperial College London, November 19, 2019, imperial.ac.uk/news/193993/ayahuasca-compound-changes-brainwaves-vivid-waking-dream/.

they alone had the training to grapple with demonic apparitions or interpret lucid visions: Bernd Brabec de Mori, email correspondence with author, January 14, 2022.

204 **turtle shells as Indigenous musical instruments:** Andrew Gillreath-Brown and Tanya M. Peres, "An Experimental Study of Turtle Shell Rattle Production and the Implications for Archaeofaunal Assemblages," *PLOS ONE* 13, no. 8 (August 2, 2018): e0201472, doi.org/10.1371/journal.pone.0201472.

"Turtle Shells Served as Symbolic Musical Instruments for Indigenous Cultures," ScienceDaily, September 5, 2018, sciencedaily.com/releases/2018/09/180905115706.htm.

Hermes myth of turtle shell transformed into a musical instrument: Hugh G. Evelyn-White, trans., "Hymn 4 to Hermes," Perseus Collection: Greek and Roman Materials, accessed February 1, 2022, perseus.tufts.edu/hopper/text?doc=Perseus:text:1999.01.0138:hymn=4.

205 **Indigenous peoples typically combine mind-altering plants with music:** Wade Davis, email correspondence with author, January 27, 2022.

peyote songs: Willard Rhodes, "Music of the American Indian: Navaho (Recording and Liner Notes)" (Archive of Folk Culture, Library of Congress, Recording, 1954; accompanying booklet, 1987), Recording Laboratory, Library of Congress, Washington, D.C. 20540, loc.gov/folklife/LP/AFSL41_Navajo.pdf.

mushroom ceremony of the Mazatec people: *Mushroom Ceremony of the Mazatec Indians of Mexico,* Album No. FR8975 (Huautla de Jimenez: Folkways Records, 1957), folkways-media.si.edu/liner_notes/folkways/FW08975.pdf.

music in the Iboga ceremony: Uwe Maas and Süster Strubelt, "Music in the Iboga Initiation Ceremony in Gabon: Polyrhythms Supporting a Pharmacotherapy," *Music Therapy Today* (online), vol. 4 (3), June 2003, wfmt.info/Musictherapyworld/modules/mmmagazine/issues/20030613105603/20030613112009/Maas.htm.

psilocybin and treatment-resistant depression: Robin L. Carhart-Harris et al., "Psilocybin With Psychological Support for Treatment-Resistant Depression: An Open-Label Feasibility Study," *The Lancet Psychiatry* 3, no. 7 (July 1, 2016): 619–27, doi.org/10.1016/S2215-0366(16)30065-7.

206 **music seemed to amplify the therapeutic effects of psilocybin:** Mendel Kaelen et al., "The Hidden Therapist: Evidence for a Central Role of Music in Psychedelic Therapy," *Psychopharmacology* 235, no. 2(February 1, 2018): 505–19, doi.org/10.1007/s00213-017-4820-5.

music described as "very useful" in bringing out the psychedelic reaction: A. Hoffer, "D-Lysergic Acid Diethylamide (LSD): A Review of Its Present Status," *Clinical Pharmacology & Therapeutics* 6, no. 2 (March 1965): 183–255, doi.org/10.1002/cpt196562183.

quality of the musical experience strongly predicted improvements in depression after psychedelic therapy: Kaelen et al., "The Hidden Therapist."

music helps shape the drug experience by evoking strong emotions and mental imagery: Mendel Kaelen, phone interview with author, July 11, 2019.

"profound alterations in emotion, mental imagery, and personal meaning": Frederick S. Barrett, Katrin H. Preller, and Mendel Kaelen, "Psychedelics and Music: Neuroscience and Therapeutic Implications," *International Review of Psychiatry* 30, no. 4 (July 4, 2018): 350–62, doi.org/10.1080/09540261.2018.1484342.

"Music reveals the soul": Rebecca Newman, "Tune in to a New Wavelength at This Psychedelic Sound Therapy Studio," *Evening Standard*, April 25, 2019, standard.co.uk/escapist/wellness/wavepaths-brick-lane-psychedelic-sound-therapy-studio-a4111796.html.

207 **"fractal mosaic of glow-pulses and flicker-riffs":** Simon Reynolds, *Generation Ecstasy: Into the World of Techno and Rave Culture* (New York: Routledge, 1999), 81–85.

208 **"Music is something that induces these very powerful experiences, life-changing experiences . . .":** Dr. Robin Sylvan: Music and the Brain series from the Library of Congress, interview by Steve Mencher, November 22, 2009, loc.gov/podcasts/musicandthebrain/podcast_sylvan.html.

"an encounter with the numinous": Robin Sylvan, *Traces of the Spirit: The Religious Dimensions of Popular Music* (New York: New York University Press, 2002).

"you're going to enter into an altered state of consciousness": Dr. Robin Sylvan: Music and the Brain series from the Library of Congress.

trances are on a spectrum of mental states: Pierre Flor-Henry, Yakov Shapiro, and Corine Sombrun, "Brain Changes During a Shamanic Trance: Altered Modes of Consciousness, Hemispheric Laterality, and Systemic Psychobiology," ed. Peter Walla, *Cogent Psychology* 4, no. 1 (December 31, 2017): 1313522, doi.org/10.1080/23311908.2017.1313522

trance in dentistry practice: Lance Rucker, in-person interview with author, Vancouver, Canada, May 15, 2017.

hypnotic trance detectable in brain scans: Heidi Jiang et al., "Brain Activity and Functional Connectivity Associated With Hypnosis," *Cerebral Cortex* 27, no. 8 (August 2017): 4083–93, doi.org/10.1093/cercor/bhw220.

209 **brain scanning during listening session of repetitive drumming—"a powerful method to alter consciousness":** Michael J. Hove et al., "Brain Network Reconfiguration and Perceptual Decoupling During an Absorptive State of Consciousness," *Cerebral Cortex* 26, no. 7 (July 1, 2016): 3116–24, doi.org/10.1093/cercor/bhv137.

ritual trance as humanity's oldest spiritual practice: Flor-Henry, Shapiro, and Sombrun, "Brain Changes During a Shamanic Trance."

210 **"Yes, very much so . . .":** Caution Shonhai, interview with author, Harare, Zimbabwe, February 28, 2019.

I interviewed seven professional mbira players and had informal conversations with a dozen others: In addition to interviews cited, passages on Zimbabwe are informed by conversations with musicians and individuals including: Musekiwa Chingodza, Jonathan Goredema, Kurai Mubaiwa, Ambuya Mugwagwa, Fradreck Mujuru, Patience Munjeri, Chiedza Mutamba, Florence Mutamba, Alois Mutinhiri, Joyce Warikandwa, and residents of Ubuntu Learning Village, as well as communications with Jennifer Kyker at Eastman School of Music and Erica Azim of mbira.org.

211 **mbira dates from the Later Iron Age and possibly much earlier:** Plan Shenjere-Nyabezi and Gilbert Pwiti, *Ancient Urban Assemblages and Complex Spatial and Socio-Political Organization in Iron Age Archaeological Sites From Southern Africa* (Brill, 2021), doi.org/10.1163/9789004500228_006.

Joshua Kumbani, "Music and Sound-Related Archaeological Artefacts From Southern Africa From the Last 10,000 Years," *Azania: Archaeological Research in Africa* 55, no. 2 (April 2, 2020): 217–41, doi.org/10.1080/0067270X.2020.1761686.

Hugh Tracey and the kalimba: "Hugh Tracey Founds African Musical Instruments (AMI): History," *Kalimba Magic* (blog), August 30, 2020, kalimbamagic.com/info/history/hugh-tracey-founds-african-musical-instruments-ami.

"helix-like weaving . . .": Brendan Baker, "Phantom Patterns: Martin Scherzinger on Shona Mbira Music," Afropop Worldwide, accessed January 11, 2022, afropop.org/articles/phantom-patterns-martin-scherzinger-on-shona-mbira-music.

the crafting and playing of mbiras inscribed in UNESCO's list of the Intangible Cultural Heritage of Humanity: "Art of Crafting and Playing Mbira/Sansi, the Finger-Plucking Traditional Musical Instrument in Malawi and Zimbabwe," UNESCO, 2020, ich.unesco.org/en/RL/art-of-crafting-and-playing-mbira-sansi-the-finger-plucking-traditional-musical-instrument-in-malawi-and-zimbabwe-01541.

212 **the value in honoring earlier generations who helped shape the current iterations of "you" and "me":** Rituals to commemorate ancestors are common in many nations, including Mexico's Day of the Dead. Fostering a sense of intergenerational connection may enhance our well-being; in a study of American children and teens, the more kids knew about their family history, the stronger their sense of control over their lives and the better their emotional health, even after traumatic events such as the September 11 attacks:

Marshall P. Duke, Amber Lazarus, and Robyn Fivush, "Knowledge of Family History as a Clinically Useful Index of Psychological Well-Being and Prognosis: A Brief Report," *Psychotherapy: Theory, Research, Practice, Training* 45, no. 2 (2008): 268–72, doi.org/10.1037/0033-3204.45.2.268.

213 **mbira denounced by some Christian groups as an instrument of the devil:** Paul Berliner, *The Soul of Mbira: Music and Traditions of the Shona People of Zimbabwe* (Chicago: University of Chicago Press, 1993).

214 **Jonathan Goredema's healing practices:** Jonathan Goredema, in-person communication with author, Ubuntu Learning Village, Gutu, Zimbabwe, February 25, 2019.

215 **"This tightness is not your pain ...":** Alois Mutinhiri, in-person translation for author, Ubuntu Learning Village, Gutu, Zimbabwe, February 25, 2019.

217 **A mid-twentieth-century survey of 488 distinct societies found that 74 percent had practices involving trance or spirit possession:** Erika Bourguignon, "A Cross-Cultural Study of Dissociational States," Ohio State University Research Foundation, Columbus, OH, 1968.

Spirit possession is most often driven by music: "Spirit Possession: An Overview," Encyclopedia.com, accessed January 11, 2022, encyclopedia.com/environment/encyclopedias-almanacs-transcripts-and-maps/spirit-possession-overview.

218 **"swing them about as if they were children":** Wade Davis, *The Serpent and the Rainbow* (New York: Touchstone, 1997), 49–50.

"there is something profoundly disturbing about spirit possession": Wade Davis, *The Clouded Leopard: Travels to Landscapes of Spirit and Desire* (Vancouver: Douglas & McIntyre, 1998), 103.

Newberg's method to study Pentecostals "speaking in tongues": Newberg recorded brain activity by injecting a small amount of a radioactive tracer before and after his subjects began dancing ecstatically and speaking in tongues. The tracer, which follows blood flow, offered a snapshot of activity at a moment in time using a technique called single photon emission computed tomography (SPECT):

Andrew B. Newberg et al., "The Measurement of Regional Cerebral Blood Flow During Glossolalia: A Preliminary SPECT Study," *Psychiatry Research: Neuroimaging* 148, no. 1 (November 2006): 67–71, doi.org/10.1016/j.pscychresns.2006.07.001.

The finding mirrored the Pentecostals' experience of their conscious selves being "taken over" by the divine spirit: "Language Center of the Brain Is Not Under the Control of Subjects Who Speak in Tongues," Penn Medicine News, accessed January 11, 2022, pennmedicine.org/news/news-releases/2006/october/language-center-of-the-brain-i.

brain imaging of Sufi mystics performing the devotional ritual Dhikr, in "mystical union with the divine": Andrew B. Newberg et al., "A Case Series Study of the Neurophysiological Effects of Altered States of Mind During Intense Islamic Prayer," *Journal of Physiology-Paris* 109, nos. 4–6 (December 2015): 214–20, doi.org/10.1016/j.jphysparis.2015.08.001.

skeptics of interpretations from brain scans of spiritual experience: Lynne Blumberg, "What Happens to the Brain During Spiritual Experiences?," *The Atlantic*, June 5, 2014, theatlantic.com/health/archive/2014/06/what-happens-to-brains-during-spiritual-experiences/361882/.

219 **Intense experiences of self-transcendence have been linked to altruistic behaviors that may last for months:** David Bryce Yaden et al., "The Varieties of Self-Transcendent Experience," *Review of General Psychology* 21, no. 2 (June 2017): 143–60, doi.org/10.1037/gpr0000102.

"spiritual practices are more likely to offer transcendent experiences in people who fully engage with them ...": Andrew Newberg, email correspondence with author, January 24, 2022.

"I am never alone": Moyo Rainos Mutamba, discussion with author, July 29, 2015.

Paul Cézanne painted Sainte-Victoire more than eighty times: "Mount Sainte-Victoire," Cleveland Museum of Art, October 30, 2018, clevelandart.org/art/1958.21.

more than half a dozen species of singing cicadas in Provence: "Enquête Cigales," Observatoire Naturaliste des Écosystèmes Méditerranéens, accessed January 10, 2022, onem-france.org/cigales/wakka.php.

220 **cicada chirping, sound volumes, loudest insect**: Patricia Bauer, "Why Are Cicadas So Noisy?," Britannica.com,accessed January 10, 2022, britannica.com/story/why-are-cicadas-so-noisy.

Diana Marques and Oscar Santamarina, "After 17 Years, Cicada Choruses Are Back," *National Geographic*, May 11, 2021, nationalgeographic.com/animals/article/after-17-years-the-cicada-choruses-are-back.

"Loudest Insect," *Guinness World Records*, accessed January 10, 2022, guinnessworldrecords.com/world-records/70619-loudest-insect.

"What Noises Cause Hearing Loss?," CDC.gov, October 7, 2019, cdc.gov/nceh/hearing_loss/what_noises_cause_hearing_loss.html.

ancient Greeks wore golden cicadas in their hair: Elwira Kaczyńska and Krzysztof Tomasz Witczak, "Cicadas in the Hesychian Lexicon," *Graeco-Latina Brunensia* 24, no. 2 (2019): 53–65, doi.org/10.5817/GLB2019-2-4.

cicadas as a symbol of Apollo, god of music and the sun: John Golding Myers, *Insect Singers: A Natural History of the Cicadas* (London: George Routledge and Sons, Ltd., 1929), 12.

Lou souleù mi fa canta (The sun makes me sing): "Les Correspondants de Frédéric Mistral," *Revue de Provence* 8ème année (1906): 91.

fable of the ant and the cicada: Francisco Rodríguez Adrados and Gert-Jan van Dijk, *History of the Graeco-Latin Fable Volume III.* (Leiden: BRILL, 2002), 146.

221 **"Human sounds must fit into and around the callings of nature . . ."**: David Rothenberg, "The Sound of Cicadas Is Music If You Take the Time to Listen," *The New York Times*, May 8, 2021, sec. Opinion, nytimes.com/interactive/2021/05/08/opinion/cicada-2021-sound.html.

"We behave like people who have right-hemisphere damage": Ideas, CBC, "Neuroscientist Argues the Left Side of Our Brains Have Taken Over Our Minds," CBC, October 22, 2021, cbc.ca/radio/ideas/neuroscientist-argues-the-left-side-of-our-brains-have-taken-over-our-minds-1.6219688.

On a walk through the forest . . .: Antonia Filmer, "Dr. Iain McGilchrist on Psychology and Neuroscience," *The Sunday Guardian Live* (blog), accessed January 10, 2022, sundayguardianlive.com/opinion/dr-iain-mcgilchrist-psychology-neuroscience.

222 **"It's not about thinking versus feeling but about two kinds of thinking"**: Iain McGilchrist, "Split Brain, Split Views: Debating Iain McGilchrist," *Pandaemonium* (blog), February 24, 2013, kenanmalik.com/2013/02/24/split-brain-split-views-debating-iain-mcgilchrist/.

"a sense of beauty and a sense of awe": Iain McGilchrist, *The Master and His Emissary: The Divided Brain and the Making of the Western World* (New Haven, CT: Yale University Press, 2019), 124.

223 **asymmetrical rhythmic groupings common in Central Asia and South Asia:** Theodore Levin, Saida Daukeyeva, and Elmira Köchümkulova, *The Music of Central Asia* (Indiana University Press, 2016).

Bruno Nettl et al., eds., *The Garland Encyclopedia of World Music*, v. 2, 5, 8: Garland Reference Library of the Humanities, v. 1169, 1191, 1193 (New York: Garland Publishing, 1998), 113.

"so unsettled, so imbalanced"; "odd times call for odd measures": Alexandra Jai, in-person communication in hand-drumming lesson, Vancouver, Canada, August 27, 2020.

224 **"irrational numbers":** The Editors of Encyclopaedia Britannica, "Irrational Number," Britannica.com, April 18, 2017, britannica.com/science/irrational-number.

ekstasis—"to stand outside one's self": The Editors of Encyclopaedia Britannica, "Ecstasy," Britannica.com, October 7, 2019, britannica.com/topic/ecstasy-religion.

225 **ruby-slipper moment:** By "ruby slippers," I mean the magic shoes in the 1939 film *The Wizard of Oz* (Victor Fleming, director; Metro-Goldwyn-Mayer, Inc.), which transported Dorothy back to Kansas from the Land of Oz when she clicked her heels and chanted the words, "There's no place like home." At the end of the film, Dorothy realizes that her home was with her all along.

226 **"The human mind makes music from sound":** Péter Sárosi, "Music and Psychedelics—an Interview With Mendel Kaelen," Drug Reporter, March 23, 2017, drogriporter.hu/en/mendel_interview/.

Coda

228 **"Music is enough for a whole lifetime, but a lifetime is not enough for music":** Sergei Bertensson, *Sergei Rachmaninoff: A Lifetime in Music* (San Francisco: Muriwai Books, 2017), public.ebookcentral.proquest.com/choice/publicfullrecord.aspx?p=4837923.

INDEX